IMPLEMENTING ENVIRONMENTAL MANAGEMENT — Policy and Techniques

Edited by David Shillito

INSTITUTION OF CHEMICAL ENGINEERS

Opinions expressed in this volume are those of
the individual authors and not necessarily those
of the Institution of Chemical Engineers.

All rights reserved. No part of this publication
may be reproduced, stored in a retrieval system,
or transmitted, in any form or by any means,
electronic, mechanical, photocopying, recording
or otherwise, without the prior permission of the
copyright owner.

**Published by
Institution of Chemical Engineers,
Davis Building,
165–171 Railway Terrace,
Rugby, Warwickshire CV21 3HQ, UK**

Copyright © 1994
Institution of Chemical Engineers
A Registered Charity

ISBN 0 85295 317 8

Cover photograph by courtesy of The British Petroleum Company plc
Other cover material by courtesy of HMSO and BSI

Printed in England by Chameleon Press Ltd, 5–25 Burr Road, London SW18 4SG

PREFACE

The objective of this book is to provide a series of snapshots of how environmental management is developing across the major companies in the process industry sector, the ideas which are being tried and the paths which are being followed. This is a period of rapid change and it is important to learn as much as possible from others — a concept known as 'bench-marking'.

Environmental management is not a new concept. It dates back to the development of the term 'Best Practicable Means' in the Alkali &c Works Acts. 'BPM' not only involved the design of the process and abatement plant but also the way in which it was maintained and operated, the way in which the plant was managed.

Of the widening of the concept to cover the complete field of environmental protection the earliest record I have found appears in the Centenary Lecture by Professor Sir Frederick Warner at the University of Newcastle in October 1971. His words on technical auditing have a very modern ring:

> 'I use a term from accountancy since it seems to me natural that progress should be made from accountability to accountancy. ... It seems to me possible that our next stage of control of environmental deterioration may well need to incorporate some form of auditing.
>
> 'I think the Public is entitled to know how far the policy of companies is directed towards obtaining a statement of ends and achievements in environmental protection which can be regularly and routinely reported on to management. There is need to know that the fine detail of operations is continuously scrutinised in order to reduce the escape of polluting material, to reuse whatever can be reused and to set targets for improvement. ... In face of all the speculation about pollution and the need to have positive programmes to ensure continuous improvement, the accumulation of data is a positive step.'

Sir Frederick's vision has taken some 20 years to mature. The first decade saw the development of environmental assessment in new projects. The import of environmental impact assessment from the USA has led progressively over the 1980s to the formalization in the EC Directive (85/337/EEC) of the regulatory requirements for assessment of environmental effects in the planning process.

'Environmental auditing' was also imported from the USA in the late 1980s. The European concept was promulgated through the International Chamber of Commerce and the ICC guide is still a valuable and concise reference. Environmental auditing is a management tool and a means to the end of environmental management.

The re-examination of safety programmes in the 1980s also led to technology transfer with environmental protection. While in the UK safety developments had concentrated on risk assessment, design reviews, hazop, etc, in the USA the ideas of gurus like Dan Petersen were being incorporated in production safety programmes. Like environmental impact assessment and environmental auditing, these ideas too had to make the journey to Europe via the multi-nationals.

The next development was truly British. While the 'quality world' has been dominated by the Japanese and Americans, the popularity of the quality assurance movement started in the UK. The progressive growth in BS 5750 and ISO 9000 started a bandwagon for management systems.

The present concept of 'environmental management systems' as specified in BS 7750: 1992 is far deeper than its quality equivalent and predecessor. It covers all the aspects of the organization, and requires assessment of environmental effects; above all it requires commitment to continual improvement. But it does not go as far as Sir Frederick forecast. It stopped short of requiring a public statement of performance.

The initial concepts for development of the EC's Eco-Audit Regulation came from environmental audit, and may possibly have predated BS 7750, but that is not really important. The evolutionary process has resulted in the new Eco-Management and Audit Regulation which requires the management system, environmental audit and the publication of an independently verified public statement.

The Regulation does not draw a line beneath the development of this arm of environmental protection. The process industries have to go further still. The next environmental bandwagon should be waste minimization. The environmental world has much more to learn from the safety world especially in design and risk assessment. Other management techniques must be adopted including the other facets of total quality management control. An excellent example is bench-marking, which is the underlying and unifying concept for this book.

David Shillito
David Shillito Associates, Consultants

David Shillito started his career with the Ever Ready Co (GB) Ltd in 1960. Then he worked in the USA and the Far East with Columbian Carbon International Inc. He joined Cremer & Warner in 1969 and at once became involved with safety and environmental protection. His main activities have been in the process engineering of air pollution control, emission abatement, trouble shooting and atmospheric dispersion. He has been involved with interdisciplinary projects of environmental assessment, incident investigation, auditing and review, and has also developed a specialist practice with odour and dust nuisance problems in both planning and litigation support. He is a member of the Institution of Chemical Engineers' Register of Safety and Environmental Professionals.

David Shillito has been closely involved with the development of safety engineering, loss prevention and risk assessment in the process industries. He has specialized in the investigation of incidents, accidents and disasters involving not only fires, explosions and crashes but environmental and product liability issues as well.

After setting up his own company in May 1991, David Shillito is now heavily involved in the management of safety and environmental protection. He maintains his specialist interest in incident investigation and legal work, especially within the field of nuisance.

The Institution of Chemical Engineers' representative on the British Standards Committees on Environmental Management Systems, BS 7750 and Environmental Auditing, he is chairman of the Institution's Editorial Panel of the *Environmental Protection Bulletin*.

PHOTOGRAPH ACKNOWLEDGEMENTS

Page 10: Elf Atochem chlorates production unit at Jarrie, France. (Reproduced by permission of Elf Atochem.)

Page 28: BASF's central power station (Ludwigshafen, Germany) with its flue gas purification system. In the middle, the two reactors of the new nitrogen-removal system can be seen; to the right of the large chimney is the desulphurization system which has been in operation since 1988. (Reproduced by permission of BASF.)

Page 34: A hazard assessment officer of the Welsh Water Authority collects a sample from the River Dee for assessment. (Reproduced by permission of AEA Technology.)

Page 48: Ciba Grimsby — production facilities and CHP. (Reproduced by permission of Ciba Grimsby.)

Page 66: National Power Eggborough. (Reproduced by permission of National Power.)

Page 80: An aerial view of the Lauterbourg site. (Reproduced by permission of Rohm and Haas.)

Page 86: Aerial view of site. (Reproduced by permission of BP Belgium.)

Page 92: The courtyard garden at British Gas Barrow Terminal. (Reproduced by permission of British Gas plc.)

Page 98: Tioxide Europe Ltd, Greatham Works, Hartlepool, Cleveland. View across Seal Sands Site of Special Scientific Interest showing seals in the foreground and the Greatham effluent discharge pipe and factory in the background. (Reproduced by permission of Tioxide Europe Ltd.)

THE AUTHORS

1. A case for environmental management
   David Shillito *(David Shillito Associates, UK)*

2. Legal aspects of environmental management and liabilities
   John Salter *(Chairman, Environmental Law Group, Denton Hall Burgin & Warrens, UK)*

3. The European Community Eco-Management and Audit Scheme
   John Stambollouian *(Environmental Protection and Industry Division, Department of the Environment, UK)*

4. The legislative requirements of integrated pollution control
   Don Munns *(Anglian Region Manager, HMIP, UK)*

5. Environmental management training — a key to the future
   Nicola Atkinson and Nigel Bell *(Centre for Environmental Technology, Imperial College, UK)*

6. Environmental management — an experience in improving environmental performance
   Andrew Sangster *(Environmental Manager, Distribution Department, Esso Petroleum Company, UK)*

7. The role of environmental audit in demonstrating improved environmental performance
   Nigel Burdett *(Manager, Corporate Environment Unit, National Power, UK)*

8. Environmental improvement through management performance standards
   Jim Whiston *(ICI Group Safety, Health and Environmental Manager, ICI, UK)*

9. Quantification of environmental management system performance as an aid to continuous improvement
   Geoff Barlow *(Assistant Regional Director for Safety, Health and Environment, Rohm and Haas (UK))*

10. Secondary products valuization
    Gilbert Devos and Walter Vissers *(BP Chemicals, Belgium)*

11. Aluminium — a case for strategic environmental management
    Charles Duff *(Director of Corporate Affairs, Norsk Hydro (UK))*

12. Environmental management in packaging — responding to the dream
    Lindsay Fortune *(Environmental Affairs Co-ordinator, Courtaulds, UK)*

CONTENTS

PREFACE   iii

THE AUTHORS   vii

1. A CASE FOR ENVIRONMENTAL MANAGEMENT   1

2. LEGAL ASPECTS OF ENVIRONMENTAL MANAGEMENT
    AND LIABILITIES   11

3. THE EUROPEAN COMMUNITY ECO-MANAGEMENT AND
    AUDIT SCHEME   21

4. THE LEGISLATIVE REQUIREMENTS OF INTEGRATED
    POLLUTION CONTROL   29

5. ENVIRONMENTAL MANAGEMENT TRAINING — A KEY
    TO THE FUTURE   35

6. ENVIRONMENTAL MANAGEMENT — AN EXPERIENCE IN
    IMPROVING ENVIRONMENTAL PERFORMANCE   49

7. THE ROLE OF ENVIRONMENTAL AUDIT IN
    DEMONSTRATING IMPROVED ENVIRONMENTAL
    PERFORMANCE   57

8. ENVIRONMENTAL IMPROVEMENT THROUGH
    MANAGEMENT PERFORMANCE STANDARDS   67

9. QUANTIFICATION OF ENVIRONMENTAL MANAGEMENT
    SYSTEM PERFORMANCE AS AN AID TO CONTINUOUS
    IMPROVEMENT   81

10. SECONDARY PRODUCTS VALUIZATION   87

| 11. | ALUMINIUM — A CASE FOR STRATEGIC ENVIRONMENTAL MANAGEMENT | 93 |
| 12. | ENVIRONMENTAL MANAGEMENT IN PACKAGING — RESPONDING TO THE DREAM | 99 |

1. A CASE FOR ENVIRONMENTAL MANAGEMENT

David Shillito

INTRODUCTION

In an ideal world there would be no need for environmental management or, for that matter, safety management. The only requirement would be for *good management*. Also, particularly in the process industries, there would be little need for regulatory control. With the high standards of the industry, commercial interests alone would ensure the success of *self-regulation*. In theory, self-regulation should not only be quite adequate but would also be a commercial necessity because of the benefits arising from:

- prevention of financial loss;
- increasing difficulty of obtaining insurance cover for environmental impairment liability;
- high cost of insurance cover;
- prevention of loss of corporate image;
- prevention of loss of confidence (usually by financial institutions);
- protection against excessive requirements of control authorities (perhaps politically based);
- protection against excessive regulation and all the hidden costs involved;
- marketing advantages.

Regulation is not simply a UK domestic problem. Changes in national regulations throughout Europe and the rest of the world can create both threats and opportunities to companies operating within international markets. Perhaps the short-term opportunities have been of greater relevance to the UK because the longer-term threats have not been perceived as tangible. Changes in regulation, at home or abroad, influence investment and costs, and thus are a destabilizing influence.

In a stable national and international economy it should, at least in theory, be possible to design and build process plants which are capable of easy and reliable operation within all the regulatory and environmental parameters which can be foreseen for the working life of the plant.

In reality the national economic situation over the last decade has not been stable enough to favour investment, for many reasons. The unpredictable markets, the high costs of investment, short-term opportunities, the slow emergence of new products, together with the 'cost reductions tendency', have all had their effects.

In many cases the lack of investment potential has effectively prolonged the lives of plants which should have been shut down years ago. To a certain degree the process industries appear to have followed the trend of other sectors and to have been slowly turning into a two class society, between the two logical extremes of:

- the dynamic organizations where development and investment has resulted in new plant, good management, with demonstrable environmental awareness, belief in self-regulation and pride in the demonstration of good performance;
- the reactive organizations where development has been limited and investment restricted, where older plants have been operated in an environment of cost reduction, and where there is talk about environmental awareness but little commitment and less involvement.

In the simplistic concepts of 'management systems' the differences which lead to this differentiation can be ascribed to 'management culture', or the adverse influences of stringent financial management for economic survival.

But the picture is not as simple as these concepts might suggest. There have been profound changes in the way in which companies have been organized and operated. The practice of decentralization in the larger companies has 'enabled' better management, or at least reduced the bureaucratic and administrative constraints to good management. It has also enabled better investment planning and the focusing of drive.

Where investment has been channelled the results have been positive. Even the larger dynamic 'first class' companies, however, can contain units which have not been targeted. These can readily develop the characteristics of the smaller 'second class' category, suffering the same inherent problems in the operation of outdated plant.

A modern integrated chemical complex is often comprised of a mixture of units of different ages in a state of 'continual improvement'. The relative luxuries of new process units sit side by side with problems left over from the past. These might be old plants, but they could also be ground contamination, poor surface drainage or constraints in services and utilities.

NEW PLANTS AND ENVIRONMENTAL MANAGEMENT

In theory, and after the sorting out of teething troubles, new process plants should be capable of operation well within any environmental constraints, whether imposed by regulation or by policy, because those factors should have been catered for in the design process. In practice, however, this situation is seldom achieved for two reasons:

- lack of technology transfer from the operations to design. Too often the new plant is treated as a project package, independent of the site as a whole. The involvement of environmental considerations in the design review stages, especially in hazop, the hazard and operability study, can help minimize these problems. Environmental problems can be built into new plants, or the new plants can generate problems in other parts of the complex;
- once operational, new plants are considered resources to be fully exploited. Most, if not all, operations groups will take pride in their ability to exceed the design expectations, with higher output levels well exceeding design capacity, high standards of reliability and compliance with regulatory requirements.

After the first year the management team is in a situation not that far removed from their counterparts operating the older plants. Environmental problems progressively start to appear. Environmental management will have to be used as a retro-fit, end-of-pipe solution to contain the excesses over the original design intentions.

OLDER PLANTS

While the management of the newer plants may be concerned with pride and record beating, the management of the older plant is more likely to be concerned with survival. How can older, less efficient plant compete against the modern counterpart? The answers again are simple:

- output levels must exceed design capacity;
- manning levels must be significantly reduced;
- maintenance costs must be reduced;
- lower specification raw materials must be considered;
- higher standards of reliability and product quality must be achieved;
- standards of emissions and discharges must meet regulatory requirements, despite the fact that these may have changed since the plant was commissioned.

Much is expected from operational management. Beside the official solutions the actual mechanisms adopted are more difficult to generalize. 'Put

good people under pressure and they will innovate'. Much has been written in management textbooks about innovation when under pressure. Innovation of course involves:
- good process engineering;
- good production engineering;
- waste minimization principles;
- good team leadership;
- ignoring the unimportant.

With good team leadership and enthusiasm, and benign inattention of higher management, the results can be quite unexpected. There is also a tendency for innovation to thrive in conditions where it is not officially recognized. This can naturally lead to unauthorized developments by individuals, working off-the-record, sometimes under the blind eye of the supervising manager. While the resulting developments may be highly productive, there can be other side effects. The relatively unimportant tasks do not get done and slide gently off the manager's desk. Typical factors perceived as being of lesser importance in the production-orientated world include:
- non-finance-related documentation of administration;
- 'little safety' — small accidents, trips, falls, cuts and burns, etc;
- process documentation, updating of changes;
- lateral communications across process units;
- documentation or notification of 'short-cuts';
- vertical communication of 'bad news'.

The innovations may well be basically very sound. The quality of the ideas may be very good, and the speed of implementation may be exceptional. What is most frequently lacking is the detailed application of the engineering practices which are advocated in all safety and environmental guidance. Hazops cost money. Re-writing and securing approval for standard operating procedures is a drag, the operators know how to run the plant anyway.

The main problem is that these deviations from good engineering practice may be tolerated because the financial rewards can far outweigh the administrative repercussions.

In the process industries there appears to be a natural conflict between official requirements for improved performance and the operational freedom to develop ways of achieving it.

There is no doubt that innovation at the production unit level is to be welcomed and supported. Full assistance must be provided to enable completion of all the associated details and the administration. Remember Rosabeth Moss Kanter's *The Change Masters*? She gave an excellent analysis of how different companies respond to this type of demand, and the ease with which the 'system' can stifle initiative.

With the older plants environmental management also has to resolve the natural conflicts between innovation and control without stifling initiative. Obviously the concept of environmental management here is not that of a quality assurance system as described in a standard, but an approach to leadership and the control of workplace behaviour.

ENVIRONMENTAL AUDITING

Environmental auditing was introduced as a management tool to help operational management. The audit concept was simply to provide feedback control to management. Several multi-national process companies adopted in-house audit programmes across their production facilities several years ago. There appears to be a consensus of opinion that these programmes have resulted not only in much improved standards of compliance but in substantial improvements in performance as well.

Most of these companies seem to have modified the purest disciplines of 'auditing practice' as set out in the 'quality' or management auditing text books. A much greater emphasis has been placed on the recognition of achievement than on the reporting of 'system non-compliance'. The benefits of constructive auditing are now preached widely. Audit disciplines have also changed to reflect the more positive aspects of the revised approach.

The success of in-house environmental audit programmes, particularly in the process industries, should be attributed to the expertise of the audit teams as well as to the company attitude and management commitment. It is easier to identify non-compliance than to recognize the advantages of innovative ideas, particularly where these might not be fully in accordance with conventional administrative wisdom. The independent in-house auditor can 'enable' the transfer of good ideas between production facilities more easily than anyone else. In some situations it is *only* the in-house auditor who can achieve this.

In-house audit programmes have had their teething-troubles. The first couple of audit cycles may be dedicated to securing the objectives of compliance. After that the success of these programmes will be self-limiting unless the scope of the activities can be extended to provide a preventative role rather than one of reaction. Experienced auditors frequently explain that the problems they find originate from within the 'management system'. Expansion of the audit programme providing a deeper examination of the management practices can sustain the effectiveness of the programme in improving performance. The experienced auditor will probably have little difficulty in accepting the validity of the need for 'management systems auditing'.

Organizations which have been slower to initiate audit programmes have attempted to catch up, usually by hiring external teams from consulting companies. Here the success may have been more limited. Certainly the audit discipline may have been just as rigorous, but the constructive element has been hindered by contractual constraints and consistency across the production facilities and between audit cycles. Unfortunately smaller organizations often cannot afford the luxury of in-house teams. This problem can be overcome by the development of a longer-term, and special, relationship with an independent consultancy. It takes time and experience to develop the constructive side of the audit programme.

Environmental audit programmes work upwards from the plant to reveal problems and solutions, to assist in the measurement of performance and in the improvement of management practices. They have tended, however, to remain centred around production facilities and have not spread to design departments, and particularly to the corporate planners. It could be argued that the success of environmental audit is due to the very practical nature of the discipline. As a bottom-up process it concentrates on what exists and what can be demonstrated, unlike the more 'woolly' world of management consultancy.

ENVIRONMENTAL MANAGEMENT SYSTEMS

The concept of 'environmental management systems' may well have arisen more from the environmental band-wagon than from industrial practice. The concept is remarkable in that it is capable of being interpreted in many different ways. Through the British Standards Institute it has been formalized as BS 7750: 1992.

BS 7750 has its origins in 'quality assurance systems' rather than in environmental auditing. Quality assurance, or quality management systems as many prefer, provides a methodology for the third party certification of compliance. BS 7750 shares this capability. Third party accreditation requires that an environmental management system be documented to a standard at which a non-specialist auditor from a certification agency can verify that the administrative processes are being fully followed. BS 7750 is thus more concerned with the documentation of the system than with its outputs

Like the quality management systems it gives the organization almost complete freedom to select environmental objectives. The minimum standard is, of course, compliance with regulatory and legislative requirements, but no company would 'knowingly' set its objectives, goals or targets below that level! The main requirements would appear to be humble objectives, complete documentation and a workforce to maintain the manuals and procedures. But it does require commitment to a policy which must include continual improvement. And commitment must start at the top.

BS 7750 has been through a pilot programme. Revisions have been suggested, a draft for public consultation has been circulated and the revised standard will probably be available in early 1994. The standard has been found to give rise to problems, particularly in defining what constitutes good environmental performance. Anyone attempting to make and complete a step change breakthrough in environmental management in less than a year must be expected to have problems in defining 'performance'.

BS 7750 has been formulated as a 'model' for good management. In this evangelical role it has followed the model of 'role culture'. It requires logical, sequential analysis, from problem definition to the identification of the solution mechanism. Planning is the essential feature. Human resources have to be planned, trained, scheduled, deployed or reshuffled like any other physical asset. In this culture belong the formal management techniques of manpower planning, assessment centres, appraisal schemes, training needs diagnosis, training courses, job rotation — in fact, all the paraphernalia of traditional management development. This may well be the right culture for the mass production industries where quality assurance is so necessary.

The problem is that there are other management cultures which work quite well, particularly in small organizations. Small organizations or organizational groups cannot afford to support the bureaucracy of the mega-system. The objectives of many management re-organizations have been to break down the mega-systems and encourage team work and the 'task culture' with its emphasis on leadership, expertise and group behaviour.

The process industries, like other capital intensive sectors, contain many types of management culture including hybrids with different cultures for different functions. Such a system may be based on a strong, but lean, line divisional structure with a 'role culture', but encourage emphasis on a team work 'task culture' in the production units, in project groups and marketing. At Board level the 'club culture' may prevail, with little sympathy for the bureaucracy below.

It is difficult to persuade such a multi-cultured organization to commit itself to a purest role culture model, as dictated by more junior bureaucrats. The operating teams may be happy with their standard operating procedures, they may welcome new editions titled 'ISO 9000, Responsible Care', but they would be horrified if they were to be lumbered with wall-to-wall manuals and procedures. Bureaucracy and innovation tend not to mix. Similarly systems designed for the auditing of administrative procedures are not good at recognizing innovation and are not well suited to transferring it to other locations in the organization. Systems auditors may even have to develop special skills.

BS 7750 can be made to fit the process industries, but it will be difficult to make the process industries fit BS 7750. What is required is a deeper vision of environmental management which will go further than environmental auditing and will be less restraining than 'enviro-assurance'. This has still to be created.

FUTURE ENVIRONMENTAL SYSTEMS

The environmental management systems concept, having come from the environmental bandwagon, is liberally provided with 'apple pie and motherhood'. Perhaps this is a necessary evolutionary stage in order to win acceptability. The current recession has helped attitudes to mature. Environmental business opportunities may have been found to be more fugitive than anticipated; the threats may also look less menacing. The cost benefit analysis of environmental management systems has frightened a range of organizations. It would appear that an air of reality has entered environmental policy making. Looking forward there appear to be four main issues to be addressed by future environmental systems:

- auditing and systems auditing, an on-going need;
- waste minimization, or enviro-loss prevention (time, materials and energy), including normal operational, unexpected and accidental wastes;

- environmental risk assessment;
- product and process development for 'sustainability'.

Auditing and systems auditing procedures for environmental management are in the process of rapid development. In the last few years a large number of companies have been exposed to the rigours of environmental auditing for the first time. The ramifications of this on strategic and tactical management decision-making are only starting to have an effect. Many companies are still in the mode of developing a reactive response to environmental legislation rather than a proactive approach to the markets that they serve. Auditing procedures provide an objective basis for assessing present environmental liabilities. Successful environmental management systems will not only assess the status quo but will also propose options for change that will bring about improvements in environmental performance. Furthermore, environmental auditing will have to become more quantitative in its approach to ensure that objective procedures are used. In many cases bench-marking will become an essential approach to the rational assessment of environmental performance.

When environmental optimization enters the world of the process engineers, environmental systems will be maturing fast. There will be a number of motivating factors that will influence the pace of implementation but in many cases commercial considerations are likely to dominate. Some companies are there already, but the vast majority of companies have still to recognize the benefits of enhanced environmental performance.

The diversity of performance in the process industries will continue to grow. The good companies will get better and the processes of natural (or economic) selection will operate. The regulators will have to control the lower ranks of the sector resulting in a reactionary management style that will further weaken their competitive position.

The next environmental wagon may well introduce further elements, perhaps even a British Standard on 'waste minimization'. If the long-term goal of environmental sustainability is to be achieved management will have to generate environmental innovation. How this is done will depend on individuals within the industry and the cultures of their organizations. Environmental management in some form or other will be essential.

2. LEGAL ASPECTS OF ENVIRONMENTAL MANAGEMENT AND LIABILITIES
John Salter

This chapter covers a wide range: the objectives of environmental management and the meaning of 'environment'; how to demonstrate environmental performance and risks arising from poor performance; the role of the environmental auditor under BS 7750 and verifier under the Eco-Management and Audit Regulation; environmental assessment of chemicals; the development of European Community standards; brief commentary on inspection, verification and accreditation; a review of auditor and manager liabilities.

LEGAL ASPECTS OF ENVIRONMENTAL MANAGEMENT
Environmental management is designed to influence the environmental performance of organizations and people. Within this concept three activities can be distinguished:
- having in place an environmental quality management system;
- regular environmental auditing or reviews;
- environmental assessment of siting, plant, plans and policies.

THE ENVIRONMENT
The expression the 'environment' embraces the built, the natural and the human environment. The built environment includes the built heritage, historic and archaeological sites, urban and rural conservation areas and the urban landscape. The natural environment includes the three important resources of air, water and soil. It includes flora and fauna and all living organisms. It includes matters relating to climatic change, temperature, rainfall and wind speeds. The human environment covers all health and safety matters — that is to say, the physical health and well being of human beings. Thus the environment covers health, safety, conservation and pollution issues, and quality control to achieve quality of life.

ENVIRONMENTAL QUALITY MANAGEMENT SYSTEMS
To achieve continuing success a company may have to demonstrate its environmental performance by securing that:

- its activities comply with applicable environmental legislation;
- its products or services are procured, produced, packaged, delivered and used and ultimately disposed of in ways which are acceptable and appropriate from an environmental viewpoint;
- its expenditure on environmental protection is prudent and cost effective;
- its strategies in planning for future investment and growth reflect market needs concerning the environment.

To avoid failure a company may have to carry out risk analysis. Why should a small or medium-sized enterprise commit resources to environmental programmes? There are certain risks arising from an unsatisfactory environmental performance in terms of legal action against the company or its directors or managers, adverse affect on image and reputation, loss of markets, inability to obtain insurance at a reasonable premium, additional costs incurred through poor utilization of resources and the general effect on personnel, customers, shareholders, bankers and regulators. From a legal point of view the likely costs arising out of poor environmental performance could include:

- fines resulting from failure to meet relevant legislative requirements. The amount of fines levied and the number of prosecutions brought have substantially increased over the last two years. Recently there have been recorded a number of instances of directors being given suspended sentences and one instance of a director being jailed. Generally speaking on summary conviction the maximum fine is £20,000 but on indictment fines are unlimited. Shell Oil, for example, was fined £1m in 1990 on indictment in a case which received much publicity for oil spillage in the Mersey;
- closing down by virtue of a prohibition notice. In Scotland the Forth River Purification Board obtained an interim injunction in January 1991 against further discharges by a company following revocation of its discharge consent;
- persons with criminal records may well not qualify to hold licences for waste transportation or disposal;
- a Compensation Order can be made under s.35 of the Powers of the Criminal Courts Act 1973 against a person found guilty of an offence if there has been non-compliance with a Nuisance Abatement Order. Clean-up costs and compensation will, in most cases, be very much more substantial than the fine. For example, a company fined £30,000 for discharging 3000 gallons of liquid waste indirectly into a sewer also had to pay £41,000 costs and compensation. Another company fined £20,000 for discharging 1000 gallons of diesel into drinking

water bore holes had to pay an estimated £500,000 costs and compensation. It should be not overlooked that private prosecutions can be brought. For example, on 31 August 1991 Greenpeace won a private prosecution against Albright & Wilson for breaches of its consent limits for zinc, chromium, copper and nickel at its works at Whitehaven in Cumbria. Whitehaven magistrates fined the company £2000 with £20,000 costs;

- the extent of possible claims and liabilities arising from adverse environmental effects and change of European standards are well illustrated by the recent case of Cambridge Water Company v Eastern Counties Leather Work plc at the time of writing on appeal to the House of Lords, involving a loss of £1.3 million;
- substantial costs can arise from legal action as regards statutory nuisance or negligence. In Halsey v Esso Petroleum [1961] 1WLR 683 the plaintiff was awarded an injunction in respect of noise from Esso's boilers, plant and vehicles and smoke/smut from its chimneys comprised in a large-scale industrial activity;
- in the event of revocation of plant operating authorizations, whether on review of otherwise, there can be substantial costs of lost production.

Legally, therefore, the advantage of having an effective environmental quality management system in place is that by the creation of customer, investor and regulator confidence, there is less chance of legal action or criminal prosecution taking place, less chance of claims arising from adverse environmental effects and less chance of authorizations being revoked or difficulties being found in raising additional bank finance. Furthermore there could be unidentified cost savings, increased business through enhanced corporate image, reduced insurance premiums and better working conditions for staff.

ENVIRONMENTAL AUDITING

Environmental auditing is the objective measurement of performance of environmental managment against standards set down by legal regulation, established criteria and company policy, with communication of findings to the company. The environmental auditor under BS 7750 has two distinct roles:

- carrying out an environmental audit to assess compliance with company policies and to assess that all sites meet regulatory requirements and applicable standards;
- checking management control of environmental practices by evaluating the quality management system in place.

The environmental verifier under the EC Eco-Management and Audit Regulation No. 1836/93 has two distinct roles:
- the verification of policies, programmes, the management system, reviews, audit procedures and cycles, the environment statement and detailed objectives for conformity with technical requirements; and checking the accuracy of the information given concerning, details of, and coverage of significant environmental issues relating to the site concerned;
- validating the environmental statement required under the regulation on technical grounds.

The regulation sets out the details of the accreditation of environmental verifiers. The verifier can either be an organization or an individual. Committee procedure has been introduced to develop guidelines for the verifiers and accreditation organizations. The change of title from Eco-Audit to Eco-Management and Audit reflects the link to BS 7750 because the scheme now has regard, not only to environmental auditing, but also to environmental management.

ENVIRONMENTAL ASSESSMENT

An assessment differs from an audit in that an assessment usually relates to proposed activities whereas an audit usually relates to existing activities. Environmental assessments are required in connection with applications for development consent under the European Directive 85/337 on the assessment of the effects of certain public and private projects on the environment and under UK regulations implementing, or at least partly implementing, this Directive. Assessments are required in connection with pollution control measures under the Environmental Protection Act 1990, for example in relation to processes for which an IPC authorization is required.

Persons notifying new substances under the Notification of New Substances Regulations 1993 (NONS93) are encouraged by Regulation 4(2) to include their own risk assessment with the notification. They certainly save an additional fee by so doing. The objective of NONS93 is to protect people and the environment from the ill effects of new substances. It requires suppliers to notify to the Health and Safety Executive (HSE) and the Department of the Environment (DOE) acting jointly their intention to place a new chemical on the market and to provide HSE, acting as the executive arm of the competent authority, with a dossier of information on the new substance. The competent authority is required to carry out a risk assessment and to make recommendations for control. Similar procedures are in place under the European Commission's Existing Substances Regulation 793/93 and underpinned by the Chemicals (Hazard Information and Packaging) Regulations 1993 (CHIP). These legal instruments provide the basis for the risk management of chemicals across the Community. HSE claims that it provides a major advance in the systematic provision of information about chemicals to both the user and the regulator and in the assessment of the risk from such chemicals.

DEVELOPMENT OF STANDARDS
The European Community has set safety standards in respect of industrial risks under the Seveso Directive 82/501. These are under review at the time of writing and a draft proposal for an amending Directive is under consideration by consultees. It has not yet been published. With regard to radioactive substances, safety standards have been set by the Euratom Directive 80/836 which again is undergoing review in the light of the recommendations made in 1990 by the International Commission on Radiation Protection (ICRP). Safety standards have also been set in respect of the manufacture of genetically modified organisms (GMOs) under two Directives 90/219 and 90/220. Safety standards in respect of chemicals have been indirectly set through risk assessment procedures under Directive 67/548 as amended in particular by the 7th Amendment Directive 92/32 and by the Existing Substances Regulation 793/93.

The establishment of an advisory committee on chemicals risk reduction has been proposed under the Fifth Action Programme. The Fifth Action Programme also envisages the development of standards for risk assessment and management generally, environmental management systems and safety management systems. Table 15 of the Fifth Environmental Action Programme sets out under the heading of 'Risk Management Targets up to the Year 2000' measures to be taken and the time frame with regard to industrial activities, chemicals control and biotechnology. The Commission is leaving to the standards institutes the development of standards of risk assessment and management and for environmental management systems. The Commission itself is reviewing the question of improved safety standards arising out of the fundamental review of Directive 82/501 and plans to report to Council and Parliament with any necessary programme of action by 1995. The Commission hopes that business enterprises utilizing the Eco-Management and Audit Scheme will set improved management and procedural standards. With regard to chemicals control, Council Regulation 793/93 regulates procedures for existing chemicals.

Hazard identification is on going with continual updating in the light of scientific and technical progress and an extension of the list of substances. Common principles for assessments are being dealt with by amendment of Directive 67/548. Risk assessment of existing chemicals is covered under Regulation 793/93 and risk assessment of some 500 active substances relating to agricultural pesticides is proposed to be covered by a Directive proposed for 1994. Legislation is planned on risk reduction programmes for 50 priority chemicals.

Development of more detailed criteria is planned for risk management for the contained use of GMOs. A regulatory instrument is proposed governing the oversight of the export of GMOs to third countries. Legislation is also proposed to regulate the safe transport of GMOs. The development of risk assessment methodologies governing common approaches and principles for environment risk assessment and common testing methods and common identification methods relating to GMOs continues.

INSPECTION
UK inspecting bodies are represented by the Council of Independent Inspecting Authorities (CIIA) established in 1973. The Council advises the UK government on proposals for EC legislation affecting safety. It has a membership of nine companies operating in the independent inspection field. Included amongst the objects of the Council are the maintenance of the highest practicable standards in the general field of inspection and certification of engineering, machinery plant and equipment. The UK representative on the European Committee of Inspection Organizations, the Confederation Européenne d'Organismes de Contrôle (CEOC) is the associated offices technical committee (AOTC) established in 1905, which is a non-profit making association providing independent advice, experience and specialist skills relating to engineering safety. It is interesting to note that AOTC has signed bilateral inspection agreements with inspection organizations in many of the EC countries permitting mutual recognition of inspection services.

CERTIFICATION
Whilst some 26 bodies have been accredited by the National Accreditation Council for Certification Bodies (NACCB) for the purposes of certifying under BS 5750 (quality management systems), no certification bodies have been formally accredited in respect of BS 7750. Some certification schemes have, however, been set up on a non-accredited basis. International ISO standards for

environmental management have not yet been adopted and so it is difficult for a certifier to know what standards are relevant.

ACCREDITATION
No accreditation bodies have yet been set up in the United Kingdom to deal with environmental management following guidelines set by the European Accreditation of Certification (EAC) and the European standards for accreditation set by CEN (EN45012).

LIABILITIES

AUDITOR LIABILITY
The early practitioners of environmental auditing developed formal audit programmes based on the principles established for financial audits. Chapter 7, entitled 'Broadening the Range of Instruments', of the Fifth Environmental Action Programme states that there will be a continuing need for legislative measures at Community level particularly in respect of the establishment of fundamental levels of environmental care and protection. This is linked to Table 17 setting out details of horizontal measures which include under the heading 'Other Economic and Market Related Instruments' proposals for the environmental audit of all major public and private enterprises. This will be achieved not only by the regulation on Eco-Management and Audit but by legislative proposals on audits to be prepared in conjunction with accountants within the time frame of the year 1994.

The profession of the environmental auditor is still in its infancy. Cases on the responsibility of financial auditors are relevant in establishing the liability of environmental auditors. In the leading case of Caparo Industries v Dickman, a well-known firm of Chartered Accountants was pursued by the purchaser of a company whose audited accounts had been prepared by that firm. It was alleged that the firm had prepared inaccurate and misleading accounts and because of this the purchaser argued that he had expended more money that he should have done in reliance upon the accounts. The House of Lords held that the accountants owed no duty at common law to the purchaser of shares as an investor or as a shareholder. Accordingly the accountants were not liable in negligence to the purchaser for any loss suffered. In another leading case James McNaughton Paper Group Limited v Hicks Anderson & Company, the court

17

held that the factors to be taken into account when considering the liability of a professional adviser include the purpose for which the report was made and communicated as well as the relationship between the parties, the state and knowledge of the auditor and the reliance placed upon the report by the person reading it.

It follows that circumstances will arise in which the environmental auditor's liability will extend beyond the duty of care owed to the client company. If, for example, the environmental auditor knew that an audit had been commissioned for the purposes of the company securing loan finance from its bankers then there may well be circumstances in which the auditor would be liable to the bank if the bank relied upon the environmental auditor's report in granting a loan. It is part of the Commission's policy to give publicity to reports. Companies, indeed, may wish to publish environmental auditors' reports as part of an annual report to shareholders, or as part of an environmental report for the benefit of the public generally, including those living in proximity to the chemical plant. The auditor should establish the purpose for which his report is being prepared.

Any communication between an environmental auditor and a solicitor in private practice will be privileged if it came into existence for the sole or dominant purpose of obtaining legal advice in relation to prospective action by a regulator. Where, however, a document has been brought into existence for a number of purposes, only one of which is related to the prospect of action by a regulator, it is only if the dominant purpose of creating the document was to use its contents to assist in the giving of legal advice or to assist in the conduct of litigation reasonably in prospect that the document will be privileged. In High-grade Traders, insurers had commissioned a report relating to a fire risk to enable insurers to make up their mind about whether to resist an insurance claim, on grounds that the insured caused the fire, and to place evidence of the cause of the fire in the hands of solicitors, if the report should suggest some probability that the fire was caused by the insured. The report was therefore prepared for two purposes. The Court of Appeal, in reversing the Judge of the Court below who held that only the second purpose was privileged, held that the two purposes were inseparable and accordingly that both were privileged. This privilege applies not only in the courts but also in the industrial tribunals. It would cover appeals, for example, against prohibition notices issued by HSE. The privilege does not extend to Public Inquiries and therefore would not cover environmental impact assessment procedures.

MANAGEMENT LIABILITY

For a company to be liable it does not have to be shown necessarily that the risk to health and safety or the risk to the environment is real. In Section 3(1) of the Health and Safety at Work etc. Act 1974 the expression 'expose to risks to their health and safety' has been said in the case of Regina v Board of Trustees of the Science Museum, Court of Appeal, Criminal Division, to imply the idea of a possibility of danger. In the Water Resources Act 1991 the term 'polluting matter' implies a discharge which is capable of causing harm. The prosecution, it was said in the case of National Rivers Authority v Egger UK Limited, Crown Court, does not have to prove that harm was actually caused.

So it is wise for a company to put in place an environmental management system with a view to the prevention of prosecution, bearing the nature of the liability in mind. Apart from criminal liability there can be liability at common law. The Cambridge Water case did not relate to any statutory provision. The dispute arose out of the common law of nuisance. Nuisance at common law is a separate but complementary source of liability to statutory nuisances arising under the Environmental Protection Act. A person suing in nuisance for damage arising out of the continuous emission of, say, fumes or leachate or smells or noise which unduly interferes with the use or enjoyment of neighbouring land may recover economic loss. The outcome is likely to be that a company would be compelled to address the issue or shut down the plant. Liability lies in the separate tort of negligence if a company commits a negligent act or omission. A previous owner of land may be liable in negligence due to physical injury or loss resulting from leaving the land in a dangerous condition. Economic loss may not, however, be recoverable under this heading.

Another common law action is trespass which is the unjustifiable intrusion on to the land of another. Direct physical intrusion and damage is required. Where a person carries on a non-natural use of land and allows something to escape from the land which causes loss to an injured person or neighbouring land, then there may be strict liability under the rule of Rylands v Fletcher, which at the present time is being reviewed by the courts, in that an appeal to the House of Lords has been lodged against the decision of the Court of Appeal in the case of Cambridge Water Company v Eastern Counties Leather plc.

INDIVIDUAL LIABILITY

Under Section 157 of the Environmental Protection Act 1990 directors, managers, secretaries and similar officers and those purporting to act in such capacity can be liable to prosecution. Charges can be brought against individuals where it can be shown that the acts committed by the body corporate were done with the consent or connivance of, or to have been attributable to any neglect on the part of those individuals. Similar wording can be found in Section 217 of the Water Resources Act 1991. In Huckerby v Elliott it was decided that a director or officer of a company consents to the commission of an offence by the company where he is well aware of what is going on and agrees to it. It also stated that there was connivance of a director when he or she had knowledge of the course of conduct likely to lead to the offence but while not actually encouraging it, said nothing about it. The section imposing criminal liability on directors also includes managers. The question of who is a manager was examined in the case of R. v Boal. It was held that the meaning of words such as 'manager' and 'officer' must be decided in the light of their context, and in that case it was held that in the criminal prosecution a manager was someone who was managing, in a governing role, the affairs of the company itself. The court said that the intention of sections like section 157 of the Environmental Protection Act was to fix with criminal liability only those who are in a position of real authority — the decision-makers within the company who have both the power and the responsibility to decide corporate policy and strategy. It is to catch those responsible for putting proper procedures in place. It is not meant to strike at underlings. It should be mentioned that under section 2 of the Company Directors' Disqualification Act 1986 the Court has power to disqualify a director for up to 15 years for an indictable offence committed with regard to (amongst other things) the management of his company or its property.

3. THE EUROPEAN COMMUNITY ECO-MANAGEMENT AND AUDIT SCHEME©

John Stambollouian

European Community (EC) environment ministers have agreed proposals for a Regulation to establish a voluntary scheme to encourage industry to undertake positive environmental management, including regular audits, and to report to the public on their environmental performance. The main features of the scheme are described in this chapter.

INTRODUCTION

At the March Council of European environment ministers agreement was reached on a Regulation to establish a voluntary Community-wide scheme to give recognition to industrial sites which follow good practice in environmental management and report regularly to the public on their environmental performance. The scheme, known as Eco-Management and Audit, is expected to come fully into operation in April 1995.

This chapter describes the objectives of the scheme and how it fits into the Government's overall environmental policy. It looks at some of the key features of the scheme and the requirements which registration will impose on company and site management. The roles of the major players in the process of registration are examined in more detail. Then follows an explanation of how the scheme will relate to the British Standards Institution (BSI) environmental management standard, BS 7750, and a brief outline of the next steps in getting the scheme into operation.

OBJECTIVES AND POLICY CONTEXT IN THE UK

The overall objective of the Regulation is to promote continuous improvements in the environmental performance of industrial activities through the promotion of:

© Crown Copyright 1993. Published with the permission of the Controller of Her Majesty's Stationery Office. The views expressed are those of the author and do not necessarily reflect the views or policy of the Department of the Environment or any other government department.

- positive environmental management;
- public disclosure of environmental impacts.

The term 'positive environmental management' means that organizations take a conscious grasp on their environmental impacts, setting themselves objectives and targets for improvement and then consciously and systematically manage their operations in order to achieve those objectives and targets.

The scheme is a voluntary one which complements legislative controls. Such statutory controls as Integrated Pollution Control and UK Local Authority Air Pollution Control set limits, imposed from outside by a regulator on pollution discharges from scheduled lists of processes. The schedules of processes chosen for controls and the limits set by the regulators are chosen on the basis of what is essential to protect the environment.

Firms wishing to participate in the Eco-Management and Audit Scheme will have to look at all the environmental impacts of their operations on site and set their own objectives and targets. Details of these and of their performance must be regularly disclosed to the public. The targets that are set will cover all significant environmental impacts and the range will be considerably wider than those which are subject to statutory controls.

The fact that a firm's targets and performance will be subject to public scrutiny is important. Unlike a quality management system where the customer specifies the quality standard required and the management system ensures that production consistently meets this standard, there is no customer/supplier relationship for environmental performance. It is therefore important that the public should be able to see and judge what the company's environmental management system is delivering. And to put pressure on companies which are not meeting reasonable targets.

The UK government does not believe that organizations will want to sign up to a voluntary scheme and report poor or indifferent performance. If they commit themselves to the scheme they will want to show that they are doing at least as well as, if not better than, their competitors. The scheme is market based and capitalizes on the fact that many leading companies now see the environment as an area of major competitive business advantage.

The scheme is designed therefore to promote positive environmental management, to provide information to the public and — through market mechanisms — to encourage organizations to compete in setting environmental improvement targets. All three of these characteristics are pillars of the UK government's environmental policy, as spelt out in the 1990 Environment White

Paper. The government has therefore strongly supported the general thrust of the Commission's proposal and has taken a very active role in developing it.

KEY FEATURES

The proposal is for a Regulation which is the most stringent form of Community instrument. It has the direct force of law in all member states without having to be transposed into national legislation. Yet it gives effect to a voluntary scheme. In this respect it exactly parallels the ecolabelling scheme which focuses on products rather than processes. The main reason for a Regulation is to ensure that the scheme comes into effect at the same time throughout the Community and operates to the same rules.

Much has been made of the possibility of this being a stalking horse for a mandatory scheme. There was no support evident during the negotiations amongst other member states for a mandatory scheme, either now or in the future. The provision in the Regulation for a review in five years' time was drafted with the intention of making clear the feeling amongst member states that consideration of the voluntary nature of the scheme should fall outside the scope of the review.

At this stage the official scheme is limited for practical reasons to the industrial sector. We in the UK have said that we intend to pilot the approach in other areas. We are already looking, with the Local Government Management Board and the Audit Commission, at how it might be applied in local authorities and it is our intention to see how it could also be applied in the private service sector.

The scheme is site based, again for essentially practical reasons. We did not want large companies to have to wait until all their sites were ready for registration before they could get ahead registering any of them. This would also have been demotivating for site managers on the best sites which were eager to join the scheme.

Why should sites wish to join the scheme? There are many reasons why companies may consider that they need to have their environmental impacts properly managed:

- it may improve the profitability of the business by revealing cost savings to be made;
- it may be that those whom the company supplies or seeks to employ become more fussy about the sort of organizations with whom they want to do business

— registration would provide a competitive advantage in the market for the best contracts and the best employees;
- it may be that those who lend money to or insure the company begin to insist upon it as reassurance that the company has its environmental impacts properly under control and knows it is complying with legislation.

Registration to the scheme confers Community-wide recognition of a site's environmental credentials, and sites which are registered to the scheme will be able to promote and publicize their involvement in a controlled way through the use of an EC symbol linked to a statement which explains its significance.

THE REQUIREMENTS FOR REGISTRATION UNDER THE SCHEME
Registration to the scheme puts certain requirements on both the site and the parent company.

Firstly, the company will need to adopt an environmental policy. This policy must provide for compliance with all environmental protection legislation and in addition it must include a commitment to continuous improvement of environmental performance.

The company must then conduct an environmental review of the activities at the site in question which should cover all the significant impacts.

In the light of the policy and review, an environmental programme is defined for the site including quantified objectives and targets. The programme must, of course, be aimed at achieving the commitments contained in the company environmental policy towards continuous improvement of environmental performance.

The sort of issues which the policy and programme would be expected to address would include:
- control and reduction of emissions;
- the management of energy and raw materials;
- waste avoidance and recycling;
- selection of new production processes and changes to existing processes;
- product planning (design, packaging, transportation, use and disposal);
- prevention, limitation and contingency plans in case of environmental accidents;
- staff information, training and external information on environmental issues.

To ensure that the policy and programme are successfully implemented there must be an environmental management system in place — systematic procedures for ensuring compliance. This management system should be integrated into the general site management practices. It can be a management system which follows a Standard or it can be a management system designed around the particular needs of the site provided it meets the requirements set out in the Regulation.

The management system must be periodically audited. The audit will provide management with information on progress in meeting the programme at the site, the effectiveness of the management system and areas where changes might be necessary.

A public statement must be produced following the initial review and at the end of each succeeding audit cycle. The statement must be designed to be read by the public. It must give an assessment of all the significant environmental issues and draw attention to key changes since the previous statement.

The statement must be validated by an independent third party — the verifier — who must also confirm that the other requirements of the Regulation have been met.

Most sites will also be required to produce simplified environmental statements annually in intervening years — that is, years where there is no completion of the audit cycle. These statements will not require validation until the end of the audit cycle.

The validated environmental statement must be sent to the Competent Body in the member state where the site is located. Once the site is registered the statement must be given suitable public exposure.

ROLES IN THE SCHEME

Mention has been made of a number of key players in the scheme. It may be helpful to look more closely at who they are and what they will do.

AUDITORS

The internal audit can be carried out either by the company's own auditors or by auditors working on the company's behalf. The important thing is that they should be sufficiently independent of the activities they audit to make an objective and impartial judgement.

ENVIRONMENTAL VERIFIERS

Environmental verifiers will usually be organizations although they could be individuals operating within a suitably limited scope. The verifier must be independent of the company and of the site's auditor. They must be accredited for the purposes of the EC Regulation.

The role of the verifier is to check that all the requirements of the Regulation have been met and to validate the statement. This implies checking the reliability of the data and information in the environmental statement and judging whether the statement adequately covers all the significant environmental issues and presents a full and fair view.

COMPETENT BODY

The role of the Competent Body is to register sites once a validated statement — and registration fee — have been received. Although the scheme is not about legislative compliance, failure to meet legislative requirements may result in the Competent Body refusing to register a site, or suspending it from the register until it is satisfied that the matter has been put right. It would have to be notified of a breach by the enforcement authority and it is not expected that minor breaches would be the subject of such notifications.

ACCREDITATION SYSTEM

The role of the Accrediting Authority will be to accredit environmental verifiers and to supervise their activities. This would take place in much the same way as the National Accreditation Council for Certification Bodies (NACCB) currently recommends the accreditation bodies for the purposes of BS 5750.

ENVIRONMENTAL MANAGEMENT STANDARDS

The Regulation provides for the European Commission to recognize national, European and international management standards as meeting the corresponding provisions in the Regulation.

It is the intention that once the Eco-Management and Audit Regulation is finally adopted, BS 7750 will be reviewed, and if necessary amended, to ensure that it is fully in line with the requirements of the Regulation.

Once that happens and the Standard is formally recognized by the Commission, sites which are certified to BS 7750 will then have met all the requirements which are covered in that Standard. These should encompass all the steps in the process of registration other than that of preparing and having validated the public statement which would be the additional requirements for registration to the EC scheme.

It can be seen therefore that the two schemes are essentially complementary.

Some firms whose business is predominantly domestic will be content to apply for BS 7750. Others who do business in the Community, or who see advantage in making a public disclosure of their environmental performance, will want to take the further step and register to the EC scheme.

The Commission intends to ask the European standardization bodies to develop and adopt Standards for environmental management systems and standards for certification, so in due course these Standards will provide another route to meeting the corresponding requirements of the Regulation.

THE WAY AHEAD

The proposals published by the Commission in March last year were the subject of intensive negotiations during the six months of the UK presidency of the Community. The text went through various amendments, and was finally adopted by Council of European Environment Ministers in June and published in the Official Journal on 10 July *(Council Regulation (EEC) No 1836/93)*.

This means that the scheme has to be fully open for business in April 1995. There is plenty to be done in the meantime by member states in establishing their competent bodies and accreditation organizations, and by the regulatory committee of member states chaired by the Commission in working out detailed guidance and codes of practice.

There is, of course, no reason why any business should wait until then to introduce an environmental management system — or indeed public reporting. Many leading companies are now doing both for many of the reasons that have been discussed. The UK government believes that this scheme will provide a further incentive for industry to do so and that it will prove popular and widely respected.

4. THE LEGISLATIVE REQUIREMENTS OF INTEGRATED POLLUTION CONTROL
Don Munns

The 1990 Environmental Protection Act (EPA)[1] brought in tough new legislation, to ensure that industries with a major potential to pollute would not only have to control emissions from their processes but would also need to consider the relative media (that is, land, sea/water or air) into which pollutants were discharged. Far greater emphasis was placed on the need to assess the effects of releases and the possible harm they might cause. Also the operation of processes was going to be far more in the public limelight, and therefore the need to manage processes properly and monitor releases was going to be of major importance.

INTRODUCTION
The Environmental Protection Act 1990 brought in sweeping changes which superseded earlier legislation under the Health and Safety At Work Etc Act 1974. These changes did away with the single media concept of pollution control and introduced IPC or integrated pollution control. Part 1 of the Act also introduced 'enabling' legislation which allowed many statutory instruments to be promulgated. It regularized the 'Polluter Pays Principal' by introducing charging for authorizations to operate. Furthermore it brought in 'freedom of information' which allowed the public, 'green' watchdogs and fellow competitors alike the opportunity to view the full application data and authorizations. Finally, the definition of harm was expanded to cover virtually every living organism on the planet. These I see as the principal changes although, of course, there were many more.

HMIP AND DEVELOPMENTS
The old HM Industrial Air Pollution Inspectorate was amalgamated with other inspectorates in 1987 to form HM Inspectorate of Pollution (HMIP), in preparation for the 1990 Act. The major changes, from the operator's point of view, were that to operate a process under the new legislation it was going to cost more money, not only in the preparation of the documentation for application and

assessing the environmental affects of the operation, but in fees and charges, planning for improvements and the not insubstantial cost of testing and monitoring the operation. Many other changes were introduced but one was to have a major impact for the regulators: the conditions under which an operator was going to be allowed to operate would be stated in the authorization, rather than be assumed under Best Practicable Means.

APPLICATION FOR AUTHORIZATION

Once an operator determines that his process is authorizable — by reference to SI 472[2] — he would need to make an application for authorization to operate.

The application procedure for authorizations laid down in the legislation (SI 507[3]) describes the range of information that must be supplied in support of the application. In the past, whereas a simple two page form was sufficient under the Act, very much more information was required. The process had to be described from an environmental viewpoint and lists of raw materials, fuels, wastes, by-products and products provided. Also all points of discharge from the process had to be identified and the concentration of substances and their mass emissions estimated. The location of the process in relation to nearby housing, buildings and sites of special scientific interest (SSSIs) was included, and the operator had to address whether his process was the best practicable environmental option (BPEO) for manufacture or production of his product or operation of his service.

The operator must also demonstrate that he is using the best available techniques not entailing excessive cost (BATNEEC) and — by means of an environmental assessment — that his operation is not harming the surroundings.

Furthermore, he must show how he plans to monitor and control his process to the regulator's satisfaction. If there are areas of his plant which are substandard or open to conjecture, then the operator must include in the application a section saying how he will investigate such areas or what his improvement plans are.

Finally he must be aware that what he puts in the application is the basis for the authorization. If it is not included, or the operator wishes to change the plant, he would need to notify HMIP before such changes or additions could be made.

To assist operators, HMIP publishes a range of guidance material[7-11]. The requirement under the legislation is for operators to provide all the necessary data but to keep it as concise and understandable as possible.

ASSESSMENT OF APPLICATION

In order for an application to be considered valid, it must not only provide the technical data described earlier[3] but also the required fee[7], based on the number of process components, and the answer to questions on confidentiality and national security.

The consultation requirements are specified in legislation (SI 507[3]) and it is important for operators to remember that if they wish to receive their authorization promptly, they must discuss their process with inspectors well in advance. The consultation process is lengthy and advertizing, statutory consultations and assessment of public comments have to be carried out rigorously. If basic information is missing or not addressed, this will only delay matters.

Finally the inspectors will make a technical assessment of the application, taking into account all comments received. Time permitting, inspectors may be prepared to submit a draft authorization to operators to ensure that it is technically correct and that nothing in the application has been misinterpreted. An authorization should be determined within four months.

AUTHORIZATION TO OPERATE

Once all the assessments have been made, it will be possible to issue an authorization. It is important for operators to realise that an authorization is complementary to the application; they both together make up the authorization to operate.

The authorization documentation will comprise a frontpiece or certificate page, followed by an explanatory memorandum and the authorized conditions. These conditions will include release limits for both concentration and mass release per annum, the testing and monitoring requirements to check the release limits, and the recording and reporting requirements. There will be some general requirements and explanations on interpretation and some specific conditions depending on the type and extent of the plant. HMIP will be particularly concerned with ensuring that the operator can demonstrate that the process is being properly controlled and monitored, especially with respect to the items mentioned above. There is also likely to be an improvement section (except possibly for new plant) which will state the details and timetable under which improvements must be made to meet new plant standards.

In some circumstances there may be issues or aspects of the plant about which there is still insufficient data for HMIP to make an informed judgement of what conditions should be applied. If this is apparent then there may be conditions set in the authorization asking for the operator to undertake a feasibility study or an investigation to supply such data. This would almost certainly result in an alteration to the authorization at some later date.

POST-AUTHORIZATION

Once the authorization has been issued and a copy placed on the public register it will be up to the operator to ensure that he adheres to the procedures described in his application and that he follows the conditions set in the authorization. He will need to test and monitor his releases from the process, to record results appropriately and report to HMIP at the stated intervals. Annual release levels will be recorded on HMIP's Chemical Release Inventory (CRI), which is a system closely followed by Ministers.

The inspector's role is now one of checking and enforcement. He may inspect the process at any reasonable time and will verify that the process is being operated in accordance with the application and authorization. He will also examine records and instrumentation and carry out check testing/monitoring of emissions. Where necessary, in the case of failing to meet authorized conditions he may take enforcement action, which may be the issue of enforcement or prohibition notices or more formal action such as prosecution.

PUBLIC REGISTERS

With the advent of the Freedom of Information legislation in Europe, the workings of the inspectorate have to be clear, concise and auditable, and papers available for public scrutiny. To meet these requirements public registers have been set up which will contain applications, public comments on consultation, the authorization documentation, monitoring data, and any enforcement data. The location of these public registers is limited at present to the three old divisional offices at Leeds, Bristol and Bedford as well as at local authority Environmental Health Departments. Under SI 507 HMIP is required to supply members of the public with copies of any of this information, at copying costs.

CONCLUSIONS

The results of this new legislation is to force industry into making a more realistic assessment of its operations and especially how these affect the environment in which we all have to live. Both operators and HMIP will be open to public scrutiny to see whether they are carrying out their responsibilities under the Environmental Protection Act and subsequent legislation. Although initially this is going to increase industrial costs, those costs will presumably be passed on to the consumer. I am sure consumers are ready to pay the price of a better and cleaner environment. HMIP is here to ensure operators are properly regulated to provide this effect.

RELEVANT LEGISLATION AND GUIDANCE

LEGISLATION
1. The Environmental Protection Act 1990.
2. The Environmental Protection (Prescribed Processes & Substances) Regulations 1991, SI 472.
3. The Environmental Protection (Applications, Appeals and Register) Regulations 1991, SI 507.
4. The Environmental Protection (Authorization of Processes) (Determination of Periods) Order 1991, SI 513.
5. The Environmental Protection (Amendment of Regulations) Regulations 1991, SI 836.
6. The Environmental Protection (Prescribed Processes & Substances) (Amendment) Regulations 1992, SI 614.

GUIDANCE
7. DOE/HMIP Fees and Charges for IPC 1993/94 (March 93).
8. HMIP Application for an Authorization under IPC.
9. HMIP Guidance Note to Applicants for Authorization — process prescribed for regulation by HMIP.
10. DOE Integrated Pollution Control — A Practical Guide — IPR/1.
11. HMIP Chief Inspectors Guidance Notes on Processes.

5. ENVIRONMENTAL MANAGEMENT TRAINING — A KEY TO THE FUTURE
Nicola Atkinson and Nigel Bell

British education stands out from that of most other countries in a number of distinct ways. Not the least of these is its progressively narrow but increasingly deep focus from early secondary school level onwards. This progressive narrowing but deepening of the curriculum continues through university even to the end-point of a PhD. One of its great advantages is the speed at which a highly qualified graduate can be produced — in around half the time it takes in many other countries. Thus traditionally the education system has been geared to the rapid production of highly qualified individuals in single and often rather narrow disciplines. It can be argued that this system, while having its merits, has fundamental weaknesses which have operated against the success of the UK economy.

Twenty years ago the concept of environmental education was scarcely heard of. The environment was viewed on the one hand as the concern of freakish pressure groups, and on the other as something that needed vague consideration by heavy industry to install the hardware necessary for reduction of pollutant discharges to meet the requirements of the regulatory authorities. For the vast bulk of business the environment was seen as irrelevant to their activities. Matters have now changed dramatically and there is reason to believe that they will continue changing at an accelerating pace into the foreseeable future. Some of the reasons for this change are examined briefly in this chapter, followed by an examination of how the higher education sector can respond to the requirements of industry for individuals trained in environmental matters.

THE NEED FOR ENVIRONMENTAL MANAGEMENT TRAINING
Courses focusing on the environment are not entirely new to this country. In fact, environmental science degrees were established in a number of universities around 20 years ago — in many cases involving more than one department. Even then, it was recognized that the environment by definition demands an inter-disciplinary approach in education which is probably best addressed

through integrated environmental science courses, such as those currently run at the Universities of Lancaster and East Anglia.

Even more recently there has been an enormous upsurge in first degree courses with the word 'environment' in the title — environmental biology, environmental engineering and so on — to the extent that there are now around 5000 undergraduates in the UK in such courses. The change has been extremely rapid. An example from Imperial College is the Mineral Resources Engineering Department in the Royal School of Mines, which formerly employed as a lecturer a botanist with interests in the environmental impacts of mining. When he left a few years ago he was not replaced. Yet now this department is mounting a new four-year course on Environmental and Earth Resources Engineering.

Whilst the recognition today of the environmental dimension of university courses is to be welcomed (Toyne Report 1993[1]), it is by no means sufficient. There is, at the same time, a separate need for courses in environmental management as such to meet the changing needs of employers. Where ten years ago there was a demand for environmental chemists and the like, today there is also a place for those within a company whose undertanding of the environment is much broader. Interdisciplinary expertise is the key. One only has to look to the legislation which is increasingly driving companies to take a greater interest in the environment to understand that this is the case.

The recently-enacted Environmental Protection Act 1990 provides a case in point. It contains two specific regimes which require companies to take a broader approach to their activities in order to ensure compliance. The first is the system of integrated pollution control which is found in Part I of the 1990 Act. It requires operators of certain 'prescribed processes' to provide detailed environmental assessments of those processes in order to obtain the authorizations needed to operate them. The conditions in those authorizations are then intended to regulate the entire process rather than just emissions into one particular medium.

In the meantime, all producers of waste now have a duty of care in respect of that waste to ensure its proper disposal. Section 34 of the 1990 Act effectively requires companies to identify their waste, determine the best route for its disposal and then take certain procedural steps designed to ensure that it reaches that site. Hefty penalties for the company and its directors await those who fail to comply with these provisions, both in the criminal and civil law.

At the same time, the law is developing other 'voluntary' schemes which are designed to assist companies to improve their environmental performance, the sanction being loss of market share. The European Community's eco-labelling and eco-management and audit schemes are both of this kind. Eco-labelling focuses upon a company's products and their environmental impacts from 'cradle to grave'. Eco-management and audit schemes look to the company's internal management systems and its environmental performance objectives. Competitors and the public are expected to place great value upon the awards granted to any company which is permitted to participate in the schemes. In each case, a new approach for compliance is required — one which is not *ad hoc* and which has the commitment of the company at the highest level.

It follows that, for a number of reasons, environmental issues are now management issues. In the first place, their resolution requires access to information within the company centrally. There will also have to be a single company policy on the question of public disclosure which may have its own legal implications. In addition, there must be a focus not just on the single components of the operation but on the whole process and the products of the company — a 'cradle to grave' appreciation of the company's activities and their impact on the environment. And there must be an ability to respond to rapid scientific and policy developments which may affect any particular point of the operation at any given time — both internally and externally. Finally, there must be a recognition of the role of organizational dynamics in successfully implementing any environmental management strategy within the company.

A company which hopes to stay ahead of environmental requirements and to take advantage of the economic opportunities offered in this sphere needs to have a thorough grasp of its whole operation and the context in which it is operating. Understanding the environmental science side of the business will not be enough — business sense will be just as important. In light of this, what are the topics which need to be addressed by future environmental managers?

Firstly, there is a continuing need to understand the technology of environmental protection, in terms of emission abatement and monitoring techniques and so on. This is the area which has traditionally been addressed by the education sector at a range of levels. What is important now is to understand the options available for the future as emissions controls are tightened, requiring completely new technologies. Furthermore, control technology should be seen in terms of the entire manufacturing process from beginning to end, aimed at operating in the most environmentally benign manner, rather than as a simple end-of-pipe solution.

Of equal, if not greater, importance now and for the future is the concept of environmental management as a whole. Training in environmental management systems is now of the highest priority, not the least with the implementation of the new standard BS 7750 and the EC eco-management and audit scheme already mentioned. Quality assurance has recently become the key to success and acceptability in many areas of the economy, and there is little doubt that the development of a framework to demonstrate compliance with stated environmental policies and objectives provides the mechanism for greatly improved environmental management, at the same time introducing a vastly increased demand for appropriate training. Coincidentally, entire new disciplines are emerging which are essential to successful environmental management, in terms of auditing, life cycle analysis, environmental insurance and the development of corporate environmental policies.

A few years ago, environmental law was a subject scarcely known outside the USA, but it has now become one of the major growth areas in the environment. Increasingly stringent legislation enacted by the EC has led as never before to the demand for appropriately trained individuals. Anticipation of future legislation and an understanding of its implications is now vital to the economic wellbeing, indeed the very survival, of many types of industry. However, these developments should not only be perceived as having a potentially negative impact, but also as presenting unparalleled opportunities in green consumerism, energy savings and markets for environmental protection products and services.

All these require an in-depth understanding of the environment from multi- and inter-disciplinary points of view. It is interesting in this respect to note the survey by Business International three years ago of 100 executives of leading European companies which showed that while 60% considered environmental issues to be significant, 32% believed they were central to the company's activities.

THE RESPONSE OF HIGHER EDUCATION

How then is the higher education sector addressing the potential market in industry for graduates with such an inter-disciplinary training? There is in fact an extremely healthy growth in first degree, postgraduate and short courses concerned with environmental issues. This is illustrated in Table 5.1 which shows the remarkable increase in numbers of first degree courses between 1988 and 1992 with the word 'environmental' in their title, offered by higher education establishments in England and Wales.

TABLE 5.1
Extract from the Toyne Report 1993

Course title	No. institutions offering courses	
	1988	1992
Environmental biology	9	13
Environmental chemistry	5	8
Environmental engineering	5	12
Environmental geography, geology, earth science	1	7
Environmental health	5	9
Environmental management, technology, monitoring, control, protection	2	22
Environmental science(s)	13	26
Environmental studies	8	9

The most prominent feature of Table 5.1 is the fact that the largest increase, both in absolute and relative terms, has taken place in courses which are categorized as covering environmental management and related topics. A fuller list of the courses which are now available in the UK can be found in the Institution of Environmental Science's *Environmental Careers Handbook*[4]. Among the more interesting developments in this field are the Confederation of British Industry (CBI) backed BA in Business Administration with Environmental Management at the Buckinghamshire College, the MSc in Environmental Diagnostics at Cranfield Institute of Technology, the MSc in European Environmental Policy and Regulation at Lancaster University, the MSc in Environmental Assessment and Management at Oxford Polytechnic and the international networking European Masters Degree in Environmental Management. The latter will be described later in this chapter.

However much training is given in the formal taught parts of undergraduate or postgraduate courses, there is no substitute for practical hands-on experience. In the case of MSc courses, ample opportunity is provided for this in the project-period which is a key component of most such courses. Projects carried out in conjunction with outside organizations on essentially real-life issues are encouraged in most MSc courses and, indeed, in part-time courses; it is often normal practice for these to be done with the student's employer. In

Imperial College's MSc in Environmental Technology, such collaborative arrangements with companies are encouraged, although there is always an internal supervisor to ensure that academic standards are maintained.

It is worth examining one of these projects in a little more detail, as it provides an excellent example of the synergy that can be achieved by a partnership in MSc training in environmental matters between industry and an academic institution. The student in this particular case had an arts background and many years experience of working in the financial sector. She coped extremely well with the MSc course, demonstrating that she was capable of understanding the scientific aspects of the environment. Her project was carried out in conjunction with 3M Health Care, who generously supported what consisted of a feasibility study of recycling asthma inhalers.

This study involved the student entering into activities completely outside her previous experience. Thus, she had to examine the distribution system for the inhalers, consult experts in materials science, talk to recycling companies and investigate the energy costs associated with the production of the inhaler and recycling of its components and translate these into potential environmental impacts in terms of pollution discharges. At the end of the day she produced a very useful report (Snow, 1992^2), but — more importantly in many ways — she developed a scheme for recycling an item which might have been considered economically impossible. Her study could well prove a model for recycling other small, widely-distributed items consisting of a range of materials.

An extremely interesting point to note is that shortly after finishing the course the student became Director of the newly-established Industry Council of Electronic Retailers, which aims to establish similar schemes for electronic components. This organization has been formed in response to German legislation requiring return of electronic components to the retailers when they are life-expired and the concern that such legislation might eventually extend to this country. Thus, this collaborative educational exercise has produced somebody at the forefront of a new environmental field of considerable concern to business.

THE INTERNATIONAL ENVIRONMENTAL DIMENSION
There is little doubt that most of the environmental pressures occurring in this country at present and in the future flow from the EC, despite talk of subsidiarity. Thus, developments in Brussels have the most profound influence on

business activities within the UK. A thorough comprehension of the underlying causes of European environmental developments and the ability to predict likely future legislation as well as to understand its likely consequences in commercial terms are arguably the most important issues to be addressed by environmental education in the 1990s.

These matters are now being addressed by the newly-inaugurated European Association for Environmental Management Education (EAEME), which has been formed as a result of a vote by the European Parliament. It is intended to become a high-profile international academic organization providing interdisciplinary postgraduate education in environmental management, with particular emphasis on the pan-European dimension.

The principal activity of EAEME is the operation of a new one-year networking Masters Degree in Environmental Management. The degree is aimed in particular at middle management in industry and other sectors in order to provide a fundamental in-depth understanding of the inter-relationships between national and international laws, environmental concerns and problems. Fourteen institutions are founder members of EAEME, being based in eight countries, including Switzerland which thus extends its remit beyond the boundaries of the EC. At present Imperial College of Science, Technology & Medicine is the only UK member of the Association.

The Masters course commences in the autumn with a four-week preparatory module, followed by a ten-week basic module held at one of several focal points. For the 1992/3 academic year, the focal points were the University of Athens; Fondation Universitaire Luxembourgeoise, Arlon in Belgium and the Université de Savoie on a campus located at Archamps in France, close to the Swiss border. The other member universities contribute to the teaching at the focal points.

The preparatory module is aimed at providing a basic understanding of the multitude of disciplines taught in later parts of the course, taking into account the very wide range of backgrounds of the students, whose previous education and experience range from law, sociology, economics and management to the natural sciences and engineering. The module includes natural sciences, environmental sciences, management science and public administration, law, economics and sociology.

The basic module builds on the first four weeks of the course and is taught by the analysis of current major environmental problems and the instruments which might be most appropriately employed for their successful resolution. It starts with two weeks on the urgent environmental issues that Europe is facing now and in the coming decades, with emphasis on transfrontier, pan-European and global issues. The next four weeks are concerned with environmental decision making, with emphasis on the organization and dynamics of the processes involved. This includes environmental legislation and regulation at national and EC level, the establishment of standards and the indicators for environmental quality, environmental impact assessment and economic instruments for environmental policy. Following this, there are two weeks on the structure and dynamics of environmental management in both public and private sectors. The basic module finishes with one week on environmental communication and education.

After completion of the basic module, the students transfer to other EAEME institutions for the remainder of the course. At this point they start a selected ten-week application module. The application modules are specialized courses which in 1993 are: legal and administrative aspects of environmental management; waste management in the market economy; environmental management in the business community; and environmental management and decision support systems.

The environmental management in the business community module has attracted the greatest interest among the students, being held at the Catholic University of Brabant, Tilburg, the Netherlands in conjunction with Imperial College. Its individual elements are taught very largely by practitioners in industry and consultancy as well as academics and these are outlined in Table 5.2. The final three months of the course consist of an individual research project which must have a strong European character and be carried out in conjunction with an appropriate outside body.

MEETING THE NEEDS OF THE FUTURE IN THE UK
While there are now a wide range of degree and diploma courses designed to produce the environmental managers of the 1990s, it must be asked whether these couses are providing the graduates required by industry and the business sectors and whether the latter have recognized the need for employees with such a training. Table 5.3 on page 44 shows the careers of the 112 graduates from the Imperial College MSc in Environmental Technology, who completed the course in 1990 and 1991.

There is a wide spread of employment across different sectors, both at home and abroad, including both public and private sectors and non-governmental

TABLE 5.2
Application module taught at the Catholic University of Brabant, Tilburg: Environmental management in the business community

Week

1. *The natural environment in the business environment*: Environmental issues from an industry perspective; the role and responsibility of industry in economic processes that cause environmental disruption and in finding solutions for environmental problems; accountability by industry; responses from industry in recent years.

2. *Environment and the business strategy*: stakeholders analysis; multinational corporate environmental policy; attitudes towards environmental issues and strategies in environmental management.

3. *Environmental management systems*: theory of environmental management; components of environmental management systems; cases for particular industries; standardization, certification; BSI pilot programme; link with quality management, health and safety.

4. *Environmental review and auditing*: history, experience in the US and in European countries; auditing theory; various types of environmental reviews and audits; methods of environmental auditing; recent developments in policy and legislation on environmental auditing; EC regulation on eco-auditing.

5. *Environmental reporting*: internal and external reporting on environmental impacts; environmental management information; structure and format of the environmental report; the environmental paragraph in the annual corporate accounts; environmental performance indicators; environmental statements and the eco-audit scheme.

6. *Life cycle analysis*: theory, methods for life cycle analysis; available models, software; data and databases on environmental profiles of materials and products, communication. Packaging, products, distribution; the German ordinance, the Dutch covenant of packaging, the coming EC directives.

7. *Eco-design*: methods for including environmental aspects in the design of products; products, processes, plants, eco-labelling, 'Blaue Engel'.

8. *Environmental marketing and communication*: Marketing of 'environmentally improved products'; communication on environmental issues with the public, residents, press and authorities.

9. *Industry involvement in environmental policy making*: actors in the process, inherent uncertainty; representation, lobbying, concerted effort; compliance, providing feed-back to regulators; environmental attitudes and business strategies on environment revisited.

10. *Management of change*: top-down and bottom-up forces in environment management; environmental management and corporate culture; the environmental management handbook and the learning organization; total process control; integration of environment, quality and worker safety; motivating people, communication with employees.

TABLE 5.3
Employment of Imperial College MSc in Environmental Technology graduates as at 27.2.92

	1990 graduates	1991 graduates
Manufacturing industry	4%	2%
Energy industry	4%	8%
International organizations (European Community, International Energy Agency, etc)	4%	2%
Non governmental organizations	16%	3%
Local government	4%	3%
UK government organizations (including regulating bodies)	10%	10%
Consultants — environmental, energy, management, engineering	27%	25%
Higher degrees/academic research	16%	21%
Overseas government organizations	2%	5%
Other	6%	10%
Not known/unemployed	8%	11%

organizations in the form of environmental pressure groups. Easily the largest employment sector is environmental consultancy of various types, including accountancy firms and management consultants — a new and welcome development. But the take-up of graduates by manufacturing industry is extremely small, and indeed is similar to that when the course started in 1977. This issue must concern academics who believe their students have much to offer this sector.

If, as a recent report seems to show, the requirement by industry is for training of current employees, then clearly the whole issue of appropriate education needs examining in some detail (David Bellamy Associates, 1993[3]). The release of an employee for a full-time one year MSc course is an extremely expensive matter and one which will not be undertaken lightly. Although this occurs sometimes, it remains the exception rather than the rule. A part-time and modular course clearly offers enormous benefits in this respect and such courses attract a ready market. There are alternatives. In particular, short courses in various aspects of environmental management are developing rapidly in many academic institutions as well as being run by other educational organizations, including commercial companies. Examples are shown in Table 5.4.

Most of these courses are of less than one week's duration and rely heavily on outside speakers with practical experience of environmental management. These developments are to be welcomed, but their impact is as yet uncertain. However, if the recommendations of a recent report commissioned by the Department for Education are acted upon by higher education institutions then some degree of environmental awareness and training is likely to become the norm for graduates entering industry. Thus, the Toyne Report has strongly recommended the incorporation of environmental issues into mainstream science and engineering degrees, while at the same time casting some doubt on the value for employers of the broader environmental science/studies undergraduate programmes.

TABLE 5.4
Examples of recent short courses on environmental issues for business

Title of course	Duration	Institution
Eco-audit and environmental management systems	2 days	Brunel University
Product life cycle analysis: practical approaches	1 day	
Presenting your product to the consumer	2 days	
Training for environmental awareness — strategies and tools	2 days	
Managing environmental change	4 days	
Environmental strategies	5 days	Imperial College
Environmental science & technology (for environmental managers and lawyers)	4 days	
EuroForum — communicating for the environment	20 days	International Centre for Conservation Education
Green Book conversations (eg, global environmental policy, Japanese environmental policy, greening politics, NGOs and environmental policy)	2 hours × 10 days	The Green College

Nine years ago, the Imperial College Centre for Environmental Technology produced a report for the European Commission on Training for Pollution Control in Industry. Using questionnaires on training requirements, it indicated very strongly the problems of smaller companies obtaining information and access to courses in pollution control — a matter of considerable concern at a time when EC environmental legislation was starting to have a major impact in the UK, and still a concern today. The major recommendation of the report was the production of distance learning packages on a wide range of aspects of pollution control and environmental management. The project was partly responsible for a Manpower Services Commission contract being obtained, jointly with Leicester Polytechnic and the University of Loughborough, to produce some 80 units under the distance-learning 'Open Tech' scheme for the training primarily of technical grade staff, but in some cases managerial personnel. This approach has a promising future.

The Government White Paper, *This Common Inheritance*, commended the Open University for its plans to strengthen its distance learning programmes in environmental issues, which has very recently produced a response from the Open University (Open University, 1992[5]). This sees the time as ripe for major initiatives in environmental education — in particular, to address the 'need for short, low-cost, high-quality modular education and training material that can be delivered quickly by a variety of means directly to the customer via existing agencies or institutions', so as to reach 'large numbers of people in a variety of roles', who would be 'serviced to achieve the necessary action outcomes'.

Another exciting initiative here is the MSc/Diploma in Environmental Management by distance learning, developed by Wye College which is aimed at students anywhere in the world. Taking this development one step further is Henley Management College's MBA which communicates via computers around the globe with its students, carrying out electronic tutorials, conferences and both course-related and social discussions. Global developments in electronic communications are bound to broaden even further the opportunities for distance learning in environmental management.

CONCLUSIONS

Training and education in environmental issues is changing direction today as never before. There is no doubt that requirements in this area have changed dramatically in the last year or so, with an increasing emphasis on environmental management in its widest sense. The future economic wellbeing of the UK requires not only a broad understanding of the ways in which environmental issues affect industry now and in the future but also — of even greater importance — the mechanisms by which the environment becomes integrated into the heart of corporate activities. To fail to address these issues will result in the UK being left further behind in the international arena.

The concern must be that time and again British industry has failed to anticipate the changing environmental world which no longer remains under the control of our national government. The future presents even greater challenges. What is the next most serious environmental issue which will hit the business world? A good case could be made for transport, which has environmental implications of the greatest significance, ranging from purely local to global. Where will the auditing of a company's environmental credentials end? Perhaps in the future a black mark may be placed against the company that uses road rather than rail transport, has widely-dispersed operations involving a large amount of transport between them, and a large fleet of company cars.

Environmental education and training are keys to the future. Environmental awareness needs to be inculcated into the population, from the individual citizen to top management. There are many ways in which this can be achieved, depending on circumstances. One of the questions facing universities is whether they should be producing specialists with an environmental training or environmentalists with a specialist training? Probably the answer is both, but with some bias towards the former! The environment is here to stay, representing not only enormous threats but also great opportunities to business. In the end, sound educational training appropriately directed will provide the key to economic progress in a sustainable and environmentally beneficial manner.

REFERENCES IN CHAPTER 5
1. Department of Education, 1993, *Environmental Responsibility. An Agenda for Further and Higher Education* (The Toyne Report), HMSO, London.
2. Snow, C, 1992, Recycling of Autohaler (TM) Inhalation Device, *MSc Thesis*, ICCET, Imperial College, London.
3. David Bellamy Associates, 1993, *Environmental Training Needs*.
4. The Institution of Environmental Sciences, 1993, *Environmental Careers Handbook*, Trotman and Company Limited, London.
5. Yoxon, M. and Blackmore, C., 1992, *The Report of the Environmental Education and Training Task Force*, Open University.

6. ENVIRONMENTAL MANAGEMENT — AN EXPERIENCE IN IMPROVING ENVIRONMENTAL PERFORMANCE

Andrew Sangster

This chapter describes how the environmental performance within one of the UK's largest petroleum distribution operations has been improved through management action with relatively small levels of investment. It covers the establishment of an environmental management system incorporating field measurements, stewardship of results and management corrective action to maintain progress towards meeting self-imposed targets. Emphasis is placed on the need to get commitment from the total workforce if lasting performance improvements are to be achieved and on the role that raising awareness plays.

INTRODUCTION

The last few years have seen a large increase in public awareness of environmental issues and concern for the way we treat the world around us. Although this has been fuelled by the media, and in some instances might have been misdirected, governments have identified with the public mood and found it necessary to respond with increasing environmental legislation, with the promise of more to come. Major industries have been quick and responsible in their turn, recognizing that public approval is needed in the long term in order to remain in business. The oil industry perhaps feels particularly vulnerable in this regard; the public wants our products but is rightly very critical when our mistakes lead to eye-catching pictures of oil-fouled birds and mammals.

At the beginning of this decade, we in Esso Petroleum felt that we needed to do more than we were already doing in order to respond to this growing environmental concern and to the consequent legislation. The oil industry has had a good safety record and is increasingly devoting management resources to ensure continuing improvement of performance. Likewise product quality has received much attention so that it now achieves consistently high levels. Environmental matters have not been ignored but have usually been seen as an offshoot of safety.

Distribution department in Esso Petroleum is responsible for transferring fuels products from refinery to customers. This involves both transportation and storage with several associated handling operations. At each stage there is potential for release of these products into the environment, either in liquid or vapour form. For many years there has been an oil loss programme in place designed to minimize stock losses and to provide a vehicle for accountability. The principal source of loss — evaporation from tankage operations — was clearly recognized many years ago, and in most cases facilities had been provided to reduce this loss as far as economically practicable.

ADVANCES IN THE EIGHTIES

Apart from tankage controls there were two other areas where distribution department had already made significant investments with consequent large reductions in hydrocarbon emissions. For many years most distribution terminals have been subject to limits on the amount of hydrocarbons that can be released in surface water discharges; consents are set by regulatory authorities (currently the National Rivers Authority (NRA) in England and Wales). All operating locations have been equipped with interceptors designed to separate free oil from the surface water run-offs and, to some extent, to remove suspended particles. Both processes rely on gravity and over time designs have been improved to give more efficient levels of performance, ensuring compliance with consent limits. Similar systems have also been installed on process water discharges such as from truck washing.

The second area of investment made in the second half of the 1980s addressed evaporative losses during trucking operations, but facilities were only installed at those locations where there was an acceptable level of economic return. Nevertheless in terms of total product losses there was a significant environmental benefit from the programme. Until this time loading of more volatile products, such as gasoline, was done through openings in the top of the tank trucks, allowing vapour to escape. When the truck then discharged its contents into customers' tanks similar volumes of vapour were again emitted from the storage venting systems. Combined losses from these operations could be as much as 0.2 to 0.3% of the total volume delivered. Closed loading systems were installed at major terminals permitting collection of the vapour and its conversion back into a liquid state. At the same time a system was developed enabling recovery of the vapour previously emitted at service stations back into the delivery truck; this vapour is then processed following reloading at the terminal. Programmes are in hand to install this system on all service station rebuilding projects and at environmentally sensitive sites.

This then was the position in early 1991; increasing legislative requirements, enhanced awareness (both public and corporate) of a need to improve environmental performance, some systems already in place and effectively reducing hydrocarbon losses to the environment, but with many smaller releases forming part of everyday operational procedure only partly or inefficiently controlled.

ENTER THE ENVIRONMENTAL GROUP
Previous experience of tackling major issues indicated that in addressing the growing number of environmental issues the best results would be obtained by setting up a dedicated resource. In the past environmental matters had always been treated as an extension of the safety group's responsibilities. The result was they had sometimes been accorded lower priorities amongst the more pressing needs of safety, and little or no expertise had been built up. In June 1991 a separate environmental group was established, initially reporting to the Operations Manager and with a mode of operation similar to, but quite separate from, the safety group.

In developing the short-term plans it quickly became apparent that the group would be starting from a very low level of knowledge about current operational performance, expectations of legislation and what could reasonably be achieved. Thus immediate priority was given to information gathering. It was known that there were some environmental controls in place in our operations and that there was a degree of monitoring, but there was no central co-ordination or assessment of performance. Some of our effluent streams had statutory consents for pollution limits and occasionally tests were carried out, but there was no follow-up action. There was already a procedure in place for disposal of oily and special waste that provided a clear audit trail. Waste disposal sites were audited but no consideration was given to whether the volumes being sent for disposal could be reduced or whether any part of them could be returned to the distribution chain.

The first steps were to ensure that every discharge was tested monthly by a competent testing agency and that the results, together with volumes of oily and special waste disposed off site during the month, were reported to the environmental group. By the end of the year there was sufficient data to give an initial indication of the level of performance in these two areas and to enable setting of realistic targets for 1992. One immediate benefit was experienced towards the end of the year when a surface water discharge at one of our terminals failed to meet the consent level during a formal NRA test. At a meeting shortly afterwards we were able, on the basis of the data we had already collected, to convince the local inspector that we knew what was happening on the site and what we had to do in order to reduce the likelihood of another failure. Our plans emphasized a high level of management attention to controlling hydrocarbon releases and their impact on the oil interceptor. Close liaison was maintained with the NRA during the remainder of the formal testing procedure, culminating in their decision not to prosecute.

IMPROVING OPERATIONAL PERFORMANCE
Setting up an environmental group proved to be the catalyst for the generation of ideas from many sources within the distribution department on how to improve operational methods and performance. Clearly this was a reflection of a general enthusiasm to make personal contributions to the environmental debate. Within two or three months it was evident that there would be a continuing large volume of performance data and ideas for improvement flowing into the group and that there was need for a mechanism to use both to best advantage. A monthly review involving the whole departmental senior management team, as part of their regular schedule of meetings, was seen as the best forum for co-ordinating the many suggestions for action and for deciding what changes were needed to achieve an improvement in performance as measured by the monthly returns.

This process subsequently developed into a more formalized environmental management system with direct feedback to each field location after every review so that performance and progress against annual targets could be shared with all staff and contractors. Early meetings were used to consider and issue instructions on 'quick fix' solutions whilst more permanent solutions were being worked out. A typical example involved the practice of using detergents to clean loading equipment and yards. Some early failures of surface water discharges to meet consent levels could be directly traced to specific cleaning operations and the subsequent entry of detergents into the interceptors, reducing their performance through emulsification of the oil present. A complete ban on the use of detergents, coupled with a switch to high pressure water washing, other than for operations discharging to foul sewer, was swiftly implemented, and made a significant contribution to reducing oil in water levels in effluent.

Other quick and cheap solutions were aimed at stopping any direct contamination of the ground during planned releases of our products and reducing the volume of hydrocarbons entering interceptors. In some instances it was merely a matter of recommissioning facilities that had fallen into disuse, because of the extra effort involved, or modifying them so that they worked more effectively.

RAISING AWARENESS
The creative eagerness in thinking up solutions has already been noted, but it was clear at the outset that it would be necessary to raise environmental awareness amongst all employees, and that all contractors should be included in the exercise, as their activities directly affect departmental environmental performance. Similar initiatives had been successful in the past to promote the message of safety in operations. The first approach was to gain the support of field managers for the plan for continuous improvement in environmental performance as laid down in the corporate objectives. This was done through management seminars in which corporate departments were invited to expound the company's policies and objectives and the environmental group had the opportunity to promote embryo plans to comply. A cascading process was used to take the appropriate messages to all field operators through their local team meetings. It was effectively reinforced by a particularly successful corporate video highlighting the environmental issues in the oil industry, what Esso had already achieved and the responsibility of individuals to make their own contribution to creating a better environment.

This initiative was rapidly followed early in 1992 by an issue of the regular departmental audio communication dedicated to environmental matters with an exposition of management policy and aims, and shortly afterwards by a roadshow which visited all major operating locations. Display material identified problem areas and floated ideas on solutions. The purpose was to stimulate thinking on how everyday operations impacted on the surrounding environment and to encourage fresh ideas on mitigation. Operators were encouraged to hold their team meetings within the show vehicle and the opportunity was taken to invite local members of regulatory bodies to view and discuss environmental issues. Since then awareness has been maintained through a poster campaign, designed to be thought-provoking, and a series of seminars for field operators addressing the attention and care needed in daily operations if environmental targets are to be met.

MANAGING THE FACILITIES BETTER

Early on each location was invited to analyse waste generation and to identify the source and volumes of special and oily waste sent off site for disposal. Results quickly showed that there were two principal sources — water draw-off and sampling operations on the main product storage tanks. Both are essential to ensure maintenance of product quality, but in each case there was potential for developing simple, relatively cheap systems for recovering the released products instead of allowing them to pass to surface water drainage systems and ultimately the interceptor. At the same time the environmental group had been looking at the availability and suitability of equipment capable of giving a high quality effluent with contamination levels significantly lower than thought possible from existing facilities. Technologies have been developed, particularly in other countries such as the United States, to meet ultra-low pollution levels permitted in discharges to watercourses. It was quickly realised that such systems would be very expensive to install and operate, particularly if all run-offs had to be treated. Also there was very limited experience in operating them. Although 'end of pipe' technology sounded attractive because it would be capable of handling any level and type of pollution, its feasibility was in doubt if not its expense.

About this time another factor became evident. From the inception of regular testing of effluent streams and centralized reporting of results, local field management had started to put in place a more positive approach to the management of oil interceptors and the quality of their effluent. Twice daily inspections were instigated and all operations involving planned release of hydrocarbons were controlled to minimize impact on the separation facilities in the drainage systems. If product releases could be deliberately kept to a minimum and then adequately diluted naturally in waste water streams, better performance could be expected from interceptors. Benefits from all this effort started to be seen in the monthly results during 1992; effluent test results improved and oily waste volumes removed from interceptors decreased.

THE RESULTS OF THE PROGRAMME

It was also becoming apparent that existing facilities could achieve the high levels of performance built into the decreasing annual performance targets, provided that operations were sensibly managed to prevent unnecessary discharges of product into the drainage streams, and some relatively small investments were

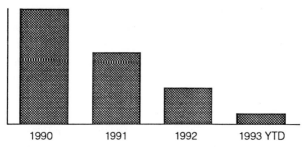

Figure 6.1 Effluent tests above target. Note: Departmental Target stricter than external Consent Levels.

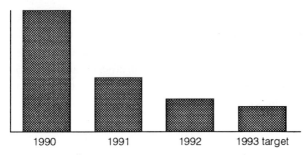

Figure 6.2 Volume of process waste.

made to assist in this control. The alternative of applying new technology to ensure attainment of the targets would be both very expensive and uncertain in outcome. The process of improving environmental performance was proving to be another example of the Pareto principle: applying management effort combined with a small amount of investment to assist that effort would lead to at least 80% of the desired results being achieved. It was also known from previous experience that solutions involving large expenditure on new equipment frequently did not achieve the expected results because the projects failed to capture the commitment of the operators.

Results to date in 1993 have shown a continuing improvement in environmental performance as targets have been progressively reduced. Figures 6.1 and 6.2 show the relative reductions in effluent test results exceeding target and the volumes of oily waste disposed off site. The environmental management system remains in place and competition between locations has been reinforced by including environmental performance in a six-monthly competition to promote excellence of operations. A programme of investment is in hand to install systems to minimize waste generation at source by capturing released products for return to the distribution chain. To date all waste from sampling operations has been eliminated and the investment has proved to be cost-effective. In parallel, small improvements have been made to some of the oil interceptors to improve their effectiveness in removing oil. The total planned expenditure of around £1.5M across twelve locations to reduce waste oil generation and thereby to improve effluent and waste disposal performance compares favourably with the need to spend a similar sum at each major location if the decision had been taken to invest in new technology to provide an 'end of pipe' solution.

Further endorsement of our policy to rely on better facilities management to achieve improved performance has come from recent discussions within the Institute of Petroleum on preparation of environmental guidelines for the industry. Optimizing the performance of existing equipment through close management attention combined with the commitment of operators is clearly seen now as the most cost-effective means of reducing the impact of petroleum distribution operations on the environment.

7. THE ROLE OF ENVIRONMENTAL AUDIT IN DEMONSTRATING IMPROVED ENVIRONMENTAL PERFORMANCE

Nigel Burdett

In the last decade or so, the relationship between industry and the environment has undergone a radical change. Companies are now modifying their practices and objectives, setting up more robust environmental management systems, including environmental audit programmes. A feature of such systems is that the audit, hitherto a fairly minor part of any management system, has often been accorded a high profile. This chapter addresses the way that the meaning of the environmental audit has changed over the last few years, with particular reference being given to National Power's activities.

NATIONAL POWER

National Power is the UK's leading electricity producer, generating over 40% of the power for England and Wales. The company operates 20 coal and oil-fired power stations and a few hydro plant, wind turbines and open cycle gas turbines. In 1992 some 44 million tonnes of coal and 1.3 million tonnes of oil were burnt to generate some 117 TWHr of electricity.

National Power has a significant impact on the environment and, as a result, we carry a great responsibility to improve our environmental performance. This is reflected in our capital investment programme for the environment. We are currently investing over £1 billion in new, cleaner, technologies as well as a further £1 billion to ensure that existing power stations manage their emissions more effectively.

£700 million is being spent in one flue gas desulphurization project to remove sulphur dioxide from our 4000 MW coal-fired power station at Drax. Three combined cycle gas turbines are currently being built and there are plans for others. We are developing port facilities to enable us to use low sulphur imported coal, closing old redundant coal-fired power stations and pursuing combined heat and power schemes. We are also involved in developing more environmentally friendly sources of power such as wind.

KEY ENVIRONMENTAL MANAGEMENT CHALLENGES

Pressures will no doubt continue for us to maintain this improvement. The scale of this expenditure has, however, encouraged us to initiate policies and develop management structures to ensure that environmental issues are more effectively integrated into our commercial policies. In other words, to ensure that the company and the environment both obtain the maximum benefit from our investments.

Our company strategy is to investigate a wide range of options for achieving both our environmental and commercial targets and then to pursue the most cost-effective solutions. We do this by:

- measuring our emissions and their impacts;
- studying the effectiveness and cost of existing and potential reduction technologies;
- developing strategies which obtain the optimum environmental benefit from our available financial resources.

Effective environmental management requires a total commitment at every level of the company and involves coupling strategic thinking and approach with practical initiatives and actions. Our goal is to generate an environmental management system which identifies where environmental risks and problems exist, devises the systems to control them and the targets to be achieved, audits performance and finally institutes corrective action where actual performance falls short of expectation.

One interpretation of environmental audit is that it is a management tool, a confidential exercise aimed at providing assurance to the company that it has environmental issues under control. This process is an integral part of the overall environmental management system which seeks to establish how well the company's management, procedures and equipment perform in delivering cost-effective compliance with legislation and company policy. It provides independent assurance that the controls which are in place are adequate to manage plant emissions, discharges and wastes.

A few years ago, the International Chamber of Commerce did sterling work to derive a definition of environmental auditing which reflects the systems orientation:

'An environmental audit is a management tool comprising the systematic, documented, periodic and objective examination of how well environmental organisation, management and equipment are performing with the aim of helping to safeguard the environment by:

(i) facilitating management control of environmental practices and
(ii) assessing compliance with company policies, which would include meeting regulatory requirements.'

But nowadays that is not sufficient. The introduction of authorizations under the Environmental Protection Act (EPA) — which require public consultation on the environmental equipment and operating practices for each plant — the draft British Standard defining the best practice in environmental management and the European Community's Eco-Management and Audit Scheme (EMAS), all point in the same direction — a need for transparency in the way we manage the environment. Companies are subject to regulation by politicians who are in turn sensitive to public opinion. Companies require investment and planning consents, operating licences and so on, all of which can be affected by the public perception of environmental issues.

We must therefore be able to demonstrate to the environmental regulatory authorities and the public, not only that we have good environmental management systems, but that we have a good case for the environmental acceptability of our existing and future plant. We believe that an open information policy is essential for this. We need to make sure that those groups with an interest in the company — including environmentalists — know our policies and what we are doing to implement them at our power stations and other sites.

Initiatives which facilitate the publication of credible information about the environmental impacts and performance of a company are now increasingly being described as environmental audits. Indeed the terminology is now so widely used that the Strategic Advisory Group on the Environment, reporting to the International Standards Organization, re-defined the environmental audit as:

'A systematic process of objectively obtaining and evaluating evidence to determine the reliability of an assertion with regard to environmental aspects of activities, events and conditions, as to how they measure to established criteria, communicating the result to the client.'

This expands the definition across the whole spectrum of business activities. At one end it is a confidential understanding of how management systems operate in the context of the company's business. At the other end, audit is an examination of actual environmental impacts and the adequacy of a company's data to satisfy the public's perceptions of those impacts.

This broadening of the definition of environmental audit has not been accidental. EMAS played a crucial part by forcing publication of the detailed environmental performance of a company. It did this by creating a linkage between the management systems audit and the external performance report. In so doing it has altered the balance between the confidential and public aspects of environmental audit and environmental information. Whilst there will always be a need for this balance to be drawn between confidentiality and disclosure, we need to recognize that the dividing line between them is moving quickly. What was considered confidential a few years ago is now freely available.

ENVIRONMENTAL MANAGEMENT IN NATIONAL POWER

We have recently re-formulated our environmental policy to respond to the increasing environmental pressures while at the same time maintaining the company's competitive position in the industry.

Our policy principles are:

- to integrate environmental factors into business decisions;
- to monitor compliance with environmental regulations and to perform better than they require, where appropriate;
- to improve environmental performance continuously;
- to review regularly at Board level, and to make public, the company's environmental performance;
- to establish a reputation for effective environmental management.

Our principles highlight the two major policy thrusts outlined above: the need for effective environmental management, building upon compliance with regulations, but doing so in a way which is transparent to the public.

Our objective is to develop a strategy in such a way that we can address both the commercial and public aspects of environment within a coherent framework. We are developing an environmental assurance system across the company to Board level, the essential feature of which is the 'ownership' of environmental responsibility by line management and the integration of environmental standards, procedures and performance reporting into normal line management processes. Each operating site and department is now establishing a robust environmental management system to control, monitor and periodically report performance.

Compliance with regulation is not sufficient. As a company we are committed to continuous improvement, particularly with regard to the way in which we manage our impacts. For this reason we have welcomed the opportunity to be involved on the pilot trial of the draft British Standard on environmental management systems (BS 7750). This is likely to provide a baseline for much of UK industry and outlines a systematic way of providing assurance of compliance with legislation and policy.

The emergent British Standard has a high credibility externally. It was also timely since it provided a sense of direction at a time when we were introducing and refining our own environmental management systems. We intend that all our major sites and departments will develop their systems with the intention of obtaining certification.

ENVIRONMENTAL AUDIT IN NATIONAL POWER
The process of evolution of the environmental audit (or as we prefer to call it, environmental review) so far in National Power has been fairly conventional. Our audits currently consist of a complete review of site compliance against legislation and corporate policy, examining the effectiveness of all environmental procedures, systems and practices against company policy and guidance. They provide information on:

- the effectiveness of the company's environmental policy;
- how well this policy is being implemented;
- the strengths and weaknesses of systems 'on the ground';
- how well staff understand environmental issues, their training and the interest they show in the environmental implications of their work;
- the methods used for collecting, evaluating and reporting environmental data;
- the site reputation and actions taken to improve it, including the management of complaints.

The exercise takes at least two man-weeks of effort on site. The aim is to give assurance to the site manager that the management system operates effectively and that non-compliancies and environmental risks have been identified and are being actively managed. It provides assistance to site management regarding the likely course of future strategy, advising on requirements for compliance with legislative and corporate standards.

The audit process has been arranged so that, in effect, the site receives environmental management consultancy from National Power staff who understand both management systems and power plant operation intimately. They also have a good view of the legislation, future company direction and ongoing environmental research programmes. They are, therefore, in a good position to advise on the development of departmental and company environmental management systems and also to shape these systems by catalysing and focusing activity.

Their aim is to strike a rapport with the power station staff and in this way to ensure that their observations and recommendations are derived from a knowledge of practicalities on the ground. The process is not one of finding fault, and the audit reports are confidential to the site manager and the audit team. The exercise has been designed to be a helpful one, and has been successful because of it.

This phase of the exercise is nearly complete. The intention is that, by the end of this year, audits will have been completed for all operating sites as well as the major headquarters functions. The aim in all cases has been to assist with the development of more effective and coherent environmental management systems. The overall cycle will have taken about two years. Audits have been undertaken of construction, closed and closing sites. We have also applied our auditing expertise to some contractors' waste disposal sites and to power stations abroad.

Drax power station. (Courtesy of National Power.)

Once effective management systems are in place, each business unit will have to audit its own management procedures, its systems and performance against target, reporting up the line as with all other line management information. The focus of the corporate environmental management systems audit will shift away from identifying problems and non-compliance areas towards assessing and improving the overall effectiveness of the management control systems. In effect this will be 'auditing the site audit'.

The credibility of an environmental audit is only as good as the people doing it. For that reason, we are sponsoring the Environmental Auditors Registration Association (EARA). This organization is currently defining the criteria of competence for an environmental auditor. We will be giving serious consideration to introducing requirements such that environmental auditors (both internal and external) operating on our sites will need to have the appropriate qualifications. Our internal auditors are already qualified to the top EARA level, that of Principal Environmental Auditor.

ENVIRONMENTAL REPORTING

We have just published our Environmental Performance Review — an open appraisal of our position. We are committed to this as an expression of openness and as a way of achieving public confidence. We believe that being reactive or secretive will simply lead to barriers, suspicion and ultimately greater regulation and control.

Contained within the Review are the principal environmental objectives for next year together with our long-term goals. To ensure that we can deliver these, we have initiated systems within the company such that each director sets performance targets for his departments which are compatible with these corporate objectives. These are monitored, reviewed, audited and reported to the Board so that we can track progress against our internal and public goals.

Our data and statements made in the Review have been independently verified and we have a commitment to further annual reports, also independently verified. In addition, we are publishing separate reports for each operating site, also using independently verified data.

The external verifiers did not repeat work carried out in our own environmental management system audits or datastream audits which provided the information for the corporate and site reviews. They verified the accuracy of the datastreams and statements made in the Review and, as far as possible, 'audited the company's internal auditors'.

Since we own several sites using similar technologies, environmental policies, audit protocols and procedures, the external verifier only needed to sample a proportion of datastreams and audit reports. For our performance report, they examined five of our sites out of a total of 21, although these represented about 50% of last year's generated output. We would not have seen any 'added value' in them seeing all of our sites.

A WAY FORWARD

Our task now is to identify the implications of future standards and build them into our current workload. We do not want to develop our environmental management systems in such a way that we have to duplicate or renew parts of them in a few years' time. When we are looking at the systems which we are developing for satisfying current compliance requirements, we should also be considering how robust they will be against the potential future demands such as BS 7750 and EMAS.

National Power is looking for a simple framework within which we can address the challenges of effective, yet publicly transparent, environmental management. Over time, the major elements of:

- a management system to BS 7750 standards;
- an internal management system environmental audit;
- an internal environmental performance audit;
- an externally verified environmental performance report;
- externally verified site reports;
- company environmental targets;

will need to come together into that framework.

In terms of system development, we are trying to identify the logic linking the requirements for EPA authorization with BS 7750 certification and EMAS. In the long run, the environmental performance report could be merged into the EMAS statement. It is certainly difficult to visualize any commercially minded company using EMAS statements *and* an environmental performance report.

We need to develop our internal environmental auditing programmes so that they can remain helpful to the site manager and provide assurance on system adequacy to the Board but, simultaneously, generate the documentation necessary to satisfy any external BS 7750 certifier or, in the longer term, an EMAS verifier.

At the heart of our thinking is the need to maximize the cost effectiveness of the whole exercise. What we have done in the preparation of this year's performance reviews is to start to learn how we could develop our environmental management systems so as to maximize the benefits yet minimize the resources required for performance review verification, for BS 7750 certification, or for EMAS verification.

CONCLUSION
National Power is attempting to develop an environmental management system which can address both the commercial and the public reporting aspects of environment. Whilst the environmental audit is a fairly small component of the total system in terms of resource allocation, the way that it develops over the next few years will be crucial in determining the shape of that system and also the credibility with which it is viewed from the outside. We look forward to contributing to that development.

8. ENVIRONMENTAL IMPROVEMENT THROUGH MANAGEMENT PERFORMANCE STANDARDS

Jim Whiston

The management of environmental systems is seen as a key area to achieve long-term, lasting improvement of chemical industry environmental performance. The ICI Group has adopted a six point improvement process. Initiated by a Board policy statement, it includes definition of performance standards, guidelines to convert these standards to locally written and implemented auditable working procedures, a system to audit these and a formal 'Letter of Assurance' to the Board of Directors from the most senior accountable manager on the degree of compliance with the standards and improvement plans to achieve full compliance.

The various rules and steps in the process are detailed in this chapter as well as arrangements for ensuring the standards reflect industry as well as company requirements.

This approach, which can be seen as a Deming 'quality improvement wheel', is a key element of ICI's response to the worldwide chemical industry 'Responsible Care' programme as well as a process to involve all senior levels of management in the achievement of environment (together with safety and occupational health) improvement throughout the company.

INTRODUCTION

The management of environmental improvement in the chemical industry is generally considered alongside that of safety and occupational hygiene since they have many common features. For small operations, it is often the same functional specialist who supports the line manager. Equally the three disciplines share common principles. In ICI, we have endeavoured to bring them together, separating only when absolutely necessary. Thus our approach to safety improvement has been extended to environment and to occupational hygiene. Here I place emphasis on environmental management but reference will be made to safety and occupational hygiene.

Environmental improvement can be considered under three separate yet interrelated elements:

- technical;

- people;
- management systems.

This is shown in Figure 8.1. However, it is important to recognize the interrelationship for the chemical industry, since if any of these elements is not addressed fully and well, then lasting long-term continuous improvement of environmental (safety and occupational hygiene) performance will not be achieved. It is most important that we never lose sight of the fact that performance improvement is our goal.

The chemical industry, despite its critics, has always given and continues to give high priority to technical improvement in environmental aspects of its plants, processes and products. Similarly, the training of staff to operate plants and processes to high environmental requirements has had high priority and provision of resource.

In this chapter I address ICI's approach to the third of the interrelated elements — Management Systems. This covers the requirement to ensure improved performance across all areas of work from research and technology through manufacturing, distribution to product safety and stewardship. Without attention to this area it is highly improbable that improvement in the other areas will be sustained over the long term.

The importance of this area of environmental improvement is demonstrated by the fact that 'Responsible Care' programmes, led and directed by a number of national chemical federations/associations, give a prominent role to implementation of management Codes of Practice. The terminology of such

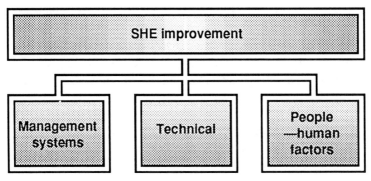

Figure 8.1 Three elements involved in environmental improvement.

Codes of Practice varies from company to company — 'management principles', or 'minimum requirements' are terms often used but within ICI we have adopted the term 'standards', carefully defined, to cover this particular area.

ICI covers a broad range of chemical businesses from pharmaceuticals, agrochemicals, explosives, speciality chemicals and paints to industrial chemicals. It is essential that there is a general approach which can encompass such diversity and also measure improvement.

The approach must cover the international dimension of differing cultures, economic and educational development. In addition it has to include arrangements to ensure that any acquired companies achieve the same performance standards as the parent company.

ICI is organized on the basis of eight international business groupings each headed by a Chief Executive Officer (CEO) reporting through a Chief Operating Officer to the Board of Directors. This business structure is supported by local subsidiaries or national companies each with a senior manager responsible for the business in that particular country. The CEO has responsibility for the overall strategic direction of the business.

Continuous improvement in environment performance through people is a key organization goal and thus any management system has to have the ability to measure performance, identify areas of improvement and ensure that improvement plans are a key feature which are monitored by those responsible for the strategic direction of the business. Within ICI, safety management systems have been part of the culture for many years but recently they have been extended to environmental and occupational health and have become part of our overall improvement process.

Here I will outline this management process, launched some time ago for safety activities and now extended to include environmental protection and occupational health. It is our firm belief that full implementation of this process will not only move us towards excellence in environmental protection (safety and occupational health) performance but also ensure compliance with the various 'Responsible Care' management Codes of Practice referred to above.

APPROACH TO ENVIRONMENTAL MANAGEMENT SYSTEMS

A six point approach has been adopted; it is outlined in Figure 8.2 on page 70. The key elements are:
- policy;

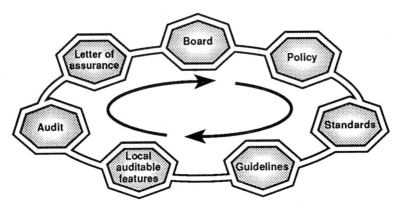

Figure 8.2   A six-point approach to environmental management.

- standards;
- guidelines;
- local auditable procedures;
- audit;
- letter of assurance.

It is important to consider each of these elements in turn as well as recognizing their interrelation.

POLICY

ICI lays down its environmental policy outlining its approach to environmental improvement as well as its arrangements for ensuring that this is carried through the company. Subsidiary companies are expected to base their policy on that of the parent and specify their arrangements for implementing the policy. Key elements from the policy are shown in Appendix 8.1 on page 76. Within the organizational arrangements, two points are of particular importance:

- the policy is set by the ICI Board and one Main Board Director is charged with overview responsibility for safety, health and environmental protection;
- the CEOs are held accountable to the Board of Directors for safety, health and environmental protection improvement. This is given the same priority as other key requirements, such as profit performance. These accountabilities are discharged through formal reporting procedures.

STANDARDS

Over recent years the key principles for safety, health and environmental protection improvement have been 'distilled' into a series of management principles which set out succinctly, in Quality Assurance terminology, the requirements and systems necessary to ensure effective implementation of the policy. These we refer to as 'standards'. Currently nineteen such standards cover these principles. They can often be regarded as the 'minimum requirements' for the company. In Appendix 8.2 on page 77 these are listed and one such typical standard is shown in detail. These standards are agreed by the CEOs and the 'Overview Director' and are then promoted throughout each business by the CEO.

GUIDELINES

It will be recognized from the standards (Appendix 8.2) that there must be more 'detail' if these are to be converted into more workable procedures to be used at all locations in the company. Since any local procedure must ensure full compliance with local legislation as well as national culture any guideline for the preparation of such a procedure can only be of a general nature and each individual location must always be allowed the opportunity to ensure that it draws from the best information available to it. Guidelines in this way can be thought of as good management practice. The nineteen standards embrace over one hundred distinct elements and a guideline has been prepared for each such element. The guidelines outline the key principles to be followed in a local procedure and give advice about how to apply the principles. Additionally they provide the references to other supporting reference data which could be relevant to the local procedures to be written. One such typical guideline is shown in Appendix 8.3 on page 78. The guidelines relevant to each standard are identified by a 'map', part of which is shown in Appendix 8.4 on page 79.

LOCAL PROCEDURES

A key principle is that procedures must be written locally to reflect national and local culture and requirements, to ensure that those who have to carry them out have a full understanding of them.

The local management or workforce prepares such procedures based on the guidelines described above. The procedures are normally written in the local language, to ensure full understanding at the working level. This is a vital stage in the process since the procedures form the basis for 'day to day', even 'hour to hour', operations.

AUDIT

Auditing is a very necessary part of the process to monitor implementation of the procedures and to identify areas for improvement.

All companies have audit procedures and within ICI three levels of audit are identified; operational, specialist and management overview.

An operational audit is a detailed systematic checking process carried out by those involved in the operation, normally the local manager and his staff. These take place frequently and routinely to check compliance with local procedures. To ensure successful performance improvement such audits must be regarded and accepted as a normal operational work feature. Identification of shortcomings, preparation and implementation of improvement plans must be the norm.

Specialist audits check that local procedures are adequate to implement the relevant standards and also address the key technical aspects of the operation — be they research and technology, manufacturing, distribution or product stewardship. These audits are carried out periodically. They will involve the local staff but will certainly include specialists in a particular discipline or area of work which is being audited. Again, identification of any defects, preparation and implementation of improvement plans are the key to success.

The third level of audit is the management overview which examines the whole management system to check that all necessary procedures are in place (including the audits referred to above) and to assess the extent to which the requirements of the nineteen standards have been achieved. It is also essential that any results of such audits must be made available to the most senior personnel (for example, CEOs). Such audits can be carried out by either a corporate function, business staff or local staff.

This flexibility recognizes the variation and variability of the businesses. However, this management audit is seen as a cornerstone of improvement and is the audit element referred to in Figure 8.2.

LETTER OF ASSURANCE

If management overview audits remain only within a local area then it is difficult to see how the CEO can be aware of the current performance within his business and how any deficiencies are being addressed. Thus within our arrangements, CEOs have to have processes in place by which the results of all management audits within their business are collated and which make them aware of both positive aspects and defects within their systems. This process is referred to as the 'Letter of Assurance' whereby the CEO has to form a view, annually, of the current status of safety, environmental protection and occupational health management systems within his business. This is based on audit findings and is summarized to the overview director. This process is now established and also includes a requirement to identify not only major defects but also the improvement plans to address any such defects so as to ensure a further move towards excellence.

The results of these Letters of Assurance are summarized in an annual report to the Board of Directors giving a view not only of performance but also of improvement plans.

BENEFITS OF THE MANAGEMENT SYSTEMS APPROACH

The benefits of the approach are that the 'Letter of Assurance' process works through each international business from a local level upwards. The process ensures that the managers of particular sections recognize the current state of environmental protection within their area of responsibility and commit themselves to the annual improvement plan. At the top of each business, the CEOs are also aware of how their responsibilities are being discharged, of the total improvement needed in their businesses and hence the resource implications.

At the top of the organization the Main Board has 'assurance' that Group safety, health and environment policies are being implemented.

The process is in essence one of plan, do, check and act — a Deming wheel.

Figure 8.3 (see page 74) shows this in a series of small improvement steps which Deming would regard as the path to 'excellence' or continuous improvement in performance through people.

PRACTICAL IMPLEMENTATION

You might now ask the question how do you carry this through every international business and site and through all territories — North, South, East and West. We have done this by bringing together the policy, the standards, the guidelines and the supporting data I have described in a book which we have titled the *ICI SHE Resource Book*.

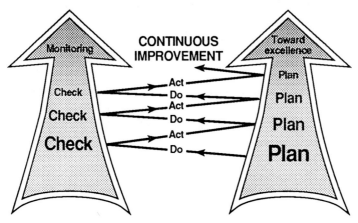

Figure 8.3   Continuous improvement.

We call this the 'White Book' and we see it having a number of significant benefits for us:

- our sites and businesses asked us to draw things together in book form so they can deliver the Board's safety, health and environment commitment internally and externally — the customer is therefore committed;
- the supporting data enables all sites and businesses to achieve what the best have been doing for some time — totally consistent with the message of total quality;
- the 'White Book' delivers ICI's pledged commitment to 'Responsible Care';
- the 'White Book' describes a unifying process applicable to all. In the fullness of time this will mean that every ICI site throughout the world will have the same base from which it has launched its safety, health and environment commitment and, more importantly, implementation;
- it recognizes that we are only as strong as the weakest of our units. It is no good, in my view, having differing standards. Implementation of our policy through our 'White Book' will ensure common standards, and common implementation taking account of compliance with specific national requirements;
- it provides the improvement 'map' for any acquisition. It enables us after any

acquisition to identify with the new management their safety, health and environment improvement plan to bring the standard of their operation up to our requirements as specified in the 'White Book'.

Thus I believe that over a period of time it will bring significant change and improvement to our already high performance.

For a company like us operating across the world it is important to recognize that languages other than English will often by required even by senior management. Thus some countries have already translated the 'White Book', or are in the process of so doing, whilst others prefer the English version. This we have left to local choice.

CONCLUSIONS

The process outlined here is being implemented across all activities of the company. It must be seen as a developing process since further areas for action will be identified and these will require the preparation of guidelines and supporting information. The process is one of continuous improvement, and can be seen as a Deming wheel of excellence. But this management process, although a key element, is only one part of environmental improvement. It involves staff from every level in the organization and thus meets one of the criteria of any continuous improvement process. Over a period of time it will bring a common approach and attitude to environmental protection — a common environmental culture, yet allows some of the businesses to reflect their specialist natures in distinctive additions so that their particular environmental performances are improved as well.

APPENDIX 8.1

ICI GROUP SAFETY HEALTH AND ENVIRONMENT POLICIES, STANDARDS AND GUIDELINES PART 1: SECTION 3

ENVIRONMENTAL POLICY IN THE ICI GROUP

It is ICI's policy to manage all of its activities so as to give benefit to society, ensuring that they meet relevant laws and regulations; that they are acceptable to the community at large; and that their environmental impact is reduced to a minimum. ICI will:

- encourage and facilitate the interchange of environmental technology throughout ICI and its subsidiary and related companies, so as to promote best practice and sustain continuous improvement in environmental performance.
- provide information and assistance to ensure that ICI's products may be used, stored and disposed of in an environmentally responsible manner.
- promote the open exchange of environmental information with customers and suppliers, as well as within ICI and the communities where ICI operates.
- provide information to enable ICI's processes, when used under licence, to be operated with acceptable environmental impacts.
- provide appropriate environmental training for employees.
- support and encourage the further worldwide development of the principles of the chemical industry's 'Responsible Care' programme and the ICC 'Charter for Sustainable Development'.
- require subsidiary and related companies to establish and implement environmental policies which accord with the policy and principles above.

IMPLEMENTATION

Compliance with regulatory legislation and standards wherever ICI operated is the minimum basis of the Group's Environmental Objectives.

ICI is committed to meeting relevant regulatory standards throughout all of its businesses worldwide.

To give greater focus to implementation of the Environmental Policy, the ICI Group has established the following Environmental Objectives:

Group Environmental Objectives

- ICI will require all its new plant to be built to standards that will meet the regulations it can reasonably anticipate in the most environmentally demanding country in which it operates that process. This will normally require the use of the best environmental practice within the Industry.
- There will be no double standards.
- ICI will reduce wastes by 50% by 1995. It will pay special attention to those which are hazardous. In addition, ICI will try to eliminate all off-site disposal of environmentally harmful effects.

APPENDIX 8.2

ICI GROUP SAFETY, HEALTH AND ENVIRONMENTAL STANDARDS

The following standards set out the essential requirements to secure implementation of the ICI Group Safety and Health, and Group Environmental Policies. They apply to all aspects of the ICI Group's activities and their implementation is mandatory. The standards are set out under the following headings:

1. Safety, Health and Environmental (SHE) Commitment
2. Management and Resources
3. Communication and Consultation
4. Training
5. Material Hazards
6. Acquisition and Divestments
7. New Plant, Equipment and Process Design
8. Modifications and Changes
9. SHE Assurance
10. Systems of Work
11. Emergency Plans
12. Contractors and Suppliers
13. Environmental Impact Assessment
14. Resource Conservation
15. Waste Management
16. Soil and Groundwater Protection
17. Product Stewardship
18. SHE Performance and Reporting
19. Auditing

A STANDARD IN DETAIL

GROUP SHE STANDARD No. 16: SOIL AND GROUNDWATER PROTECTION
'There shall be arrangements to minimise risk of comtamination of land and groundwater. Each location shall maintain a dossier which records the history and contamination of the site; and a register of all leaks and spills which may have potential for land contamination. Each location shall assess and review at regular intervals possible contamination on its land and the need for protective, containment or remediation measures.'

APPENDIX 8.3

RELEVANT GROUP DOCUMENTATION
GEP 1: General Requirements for Group Engineering Procedures
GEP 8: Protection of Ground and Groundwater
GEP 8.1: Protection of Ground and Groundwater
GEP 16.2: Procedures for the Assessment and Subsequent Management of Soil and Groundwater Contamination
ICI Group Environmental Guidelines on Groundwater Contamination Volumes 1–5.
ICI Site Investigation Protocols for Potentially Contaminated Soil and Groundwater.

SPECIFIC LEGAL REQUIREMENTS
Local procedures shall take account of relevant legislation.

PRINCIPLES TO BE FOLLOWED
(a)   Design of new installations needs to take account of the necessity to protect ground and groundwater and to ensure that any spillage or leakage of chemicals or effluent that could cause Significant Environmental Harm is prevented from reaching the Ground or stormwater Drainage.

Groundwater Monitoring boreholes should be installed on new sites.

For existing installations a strategy for improvement should be developed that ensures the engineering assets progressively achieve compliance.

When an installation is to be demolished, a record should be retained of the assets abandoned in the ground and an investigation carried out of ground and groundwater contamination.

Investigations of ground or groundwater contamination should be carried out in accordance with the site assessment procedures and the protocols for site investigation.

(b)   Drainage, Landfill, Lagoons and Bunds where a loss of conatinment could lead to Significant Environmental Harm or safety risk should be registered, periodically inspected and operated within their design limits. An engineer with the appropriate theoretical and practical knowledge and experience is required to approve design, modification and abandonment

(c)   Record drawing should be maintained of all buried Drainage and any modifications.

New Drainage forming a buried system ought not to be used for the transport of effluent when a failure could lead to Significant Environmental Harm; in such cases a pumped overground system or systems in which leakage can be monitored should be used.

When a drain is to be abandoned, an assessment should be made of the requirements for the decontamination and sealing of the drain.

APPENDIX 8.4

ICI GROUP SAFETY, HEALTH AND ENVIRONMENTAL STANDARDS AND SUPPORTING GUIDELINES

Group SHE standard (GS)		Group SHE guidelines (GG)	
GS	Subject	GG	Subject
1.	SHE commitment	1.1	Organization and arrangements to implement SHE policies
		1.2	Occupational health policy matters
2.	Management and resources	2.1	Occupational health organization
		2.2	SHE improvement plan
		2.3	Management responsibilities on sites and works for SHE matters
		2.4	Management structure for major hazard plants
		2.5	Definition of responsibility between businesses and manufacturing sites
		2.6	Provision of occupational health facilities
3.	Communication and consultation	3.1	Material hazard communication
		3.2	Communication processes
		3.3	Transfer of technology
		3.4	Public statements
4.	Training	4.1	Training
5.	Material hazards	5.1	Chemical inventory
		5.2	Hazard identification and assessment
		5.3	Exposure limits
		5.4	SHE classification, labelling and documentation for transport
		5.5	Specification for transport tanks, containers, packages, off-site storage tanks and warehouses and other distribution equipment
		5.6	Provision of SHE information to customers
		5.7	Inspection of customer bulk storage installations before first delivery
		5.8	Biological hazards
6.	Acquisitions and divestments	6.1	Procedures for divestment

9. QUANTIFICATION OF ENVIRONMENTAL MANAGEMENT SYSTEM PERFORMANCE AS AN AID TO CONTINUOUS IMPROVEMENT

Geoff Barlow

Certification of environmental management systems to a standard will become increasingly important in the next few years. The ability to be able to demonstrate continuous improvement in system performance is likely to be a common requirement. This chapter describes one approach to the problem.

BACKGROUND

Rohm and Haas Company introduced its current style of environmental audit programme in the late 1980s. The aim of the programme was to give assurance to the senior management of the company that plant operations complied with applicable regulations, company policies and best management practices. The programme operates worldwide and uses a pool of trained auditors from within the company. A series of protocols has been developed which are tailored to the regulatory requirements of each country in which we operate. This is supported by a common protocol for company policy, engineering standards and best management practice. The audit examines the environmental management systems in place at a facility which address regulatory, corporate and best management practice requirements. The audit programme has been instrumental in the development and improvement of such systems.

It was recognized during 1991, however, that certification to an environmental management system standard was going to be desirable for a number of reasons — potential and existing customers would require it, it would be helpful in communications with stakeholders and most importantly it would result in a more efficient operation. Early in 1992 the decision was taken to 'benchmark' one of our sites in Europe in terms of its environmental management systems development.

Periodically the company uses an external consultant to carry out an environmental compliance audit. The reasons for this are that it allows us to compare the results of our internal audits with those obtained by an independent external organization, it provides another level of assurance to the senior management of the company and it gives us the chance to learn any lessons there might be from a 'state-of-the-art' audit. The consultant was requested to carry out a review of the operation that would satisfy the requirements of the European Community Environmental Management and Audit Scheme (CEMAS) in addition to the routine compliance audit.

The environmental management systems adopted by a plant to deliver the requirements of the various policies depend to some extent on the size and complexity of the operations. The Lauterbourg plant in France is our largest in Europe employing some 700 people and manufacturing products for five of our core businesses. It was selected for the benchmark study because we felt that its environmental management systems were quite well developed. Based on the lessons we learned and the results of this assessment, we could then embark on a programme for the rest of our European plants.

In mid-1992 we volunteered, and were selected, for participation in the EEC pilot study for the CEMAS scheme. The review was, therefore, focused on the requirements of the regulation.

ENVIRONMENTAL MANAGEMENT SYSTEMS ASSESSMENT
Based on a comprehensive package of information, including previous audit reports, supplied by the plant to the consultant and also on a site visit, an 'issues inventory' was developed:
- general environmental management;
- hazardous materials;
- air pollution control;
- water pollution control;
- solid and hazardous waste management;
- soil and groundwater contamination;
- noise;
- energy management;
- product stewardship.

Each of the issues was then further divided into individual elements for assessment.

The assessment team consisted of consultant staff, site management, corporate audit staff, regional environmental and operating staff. This gave a balance of objectivity, local knowledge and experience.

Taking the general environmental management issue as an example, it was broken down into the following elements for assessment:
- environmental policy statement;
- awareness at facility level of management's environmental goals;
- defined responsibilities;
- staffing;
- background experience of environmental staff;
- assessment of environmental effects of new projects or modified processes;
- budgeting for environmental expenditures;
- regulatory tracking and communication at corporate level;
- regulatory tracking and communication at site level;
- compliance inspections.

Using a similar approach to that adopted in the GEMI (Global Environmental Management Initiative), publication for environmental self assessment each element was rated on a scale of 1–5.

The scale represents the various stages of management system development for a particular element. In general terms, the more formalized, documented and progressive the system was, the higher the score that would be achieved. A score of 3 was judged as satisfactory (it complied with any applicable regulatory demands and would meet the requirements of the scheme) for a medium-size multi-product plant. In total more than 60 elements of the site environmental management system were assessed in this manner.

In addition to assessing the current status of each of the elements, a target score was also determined. In most, but not all, cases this was 5. Examples of where target scores of less than 5 might be appropriate are if issues are not handled fully at site level — for instance, product stewardship is managed by the head office in Paris. Also the resources and/or cost-benefit of moving from a satisfactory score to the maximum may be judged to be unacceptable.

The average and target score for each issue was determined using a simple average — equal weight was given to each of the elements making up the issue. Greater value may be achieved in later reviews by weighting issues judged by the assessment team to be more important.

For the Lauterbourg plant the results are shown overleaf in Table 9.1.

TABLE 9.1
Results from the Lauterbourg Plant

	Current performance	Target
Hazardous materials	3.2	5.0
Air quality	3.2	5.0
Water quality	4.7	4.8
Solid and hazardous waste	4.1	4.8
Soil and groundwater	4.3	5.0
Noise	4.7	5.0
Energy	4.0	5.0
Product stewardship	4.0	5.0
General environmental management	3.9	4.9

DEVELOPING THE IMPROVEMENT PROGRAMME

Using the current and target scores for each element, the site management was then able to construct an improvement programme which reflected the priorities brought up in the review. Taking the highest priority items first, a documented programme, including a timetable for implementation, was produced. A summary of these items was then included in an environmental statement as a programme for going beyond compliance. In every case the time to reach the target level was within the audit cycle frequency, thus progress towards the achievement of goals could be confirmed during the next audit.

Clearly, substantial completion of the improvement programme should help considerably in gaining credibility for future plans. Further reviews, in conjunction with audit findings, would then be used to establish new target levels, programmes and timetables and progress may be similarly monitored by the next round of audits of the actual versus target situation.

GENERAL COMMENTS

We believe that this numerical approach meets the requirements of the regulation for a concise and non-technical presentation of environmental management system performance. It does, however, raise some concerns. The scores produced are unique to the particular audit/review and site. It would be meaningless to compare scores from one chemical plant with another. Equally, it would serve no useful purpose to try and assess a facility's improvement by comparing scores achieved in one review with those subsequently produced. The performance criteria, regulatory environment or company policy/philosophy may have changed substantially in the intervening period.

10. SECONDARY PRODUCTS VALUIZATION
Gilbert Devos and Walter Vissers

An activity has been set up within BP Chemicals (BPC) to establish applications and markets for BPC by-product waste streams. This activity is fully in line with the company's health, safety and environment policy and combines a target of environmental improvements with cost avoidance and adding value targets.

This paper focuses on:
- the mission and scope of secondary products valuization;
- the development of this activity within BPC as a multinational organization;
- the results and future potential.

INTRODUCTION
Chemical processes have a nasty property of yielding not only the desired molecule but also a number of undesired ones.

Despite all process improvements, the ideal process with 100% yield remains a utopian dream for most of our industrial chemicals.

The continuous improvement objective, which forms the cornerstone of the BPC total quality programme, tackles this area via many routes. Besides routes of improved operations, improvements in reaction chemistry, the development of alternative processes and products, BPC is approaching by-products of its chemical processes as an opportunity for creating a 'product' to suit specific markets and applications. To that purpose, a Secondary Products Valuization team (SPV team) has been formed.

Waste minimization is expensive. Companies are lucky to get a return on half of their projects. In fact managers are delighted when they find a venture that provides some payback. In most of the cases, the determining factor is 'What is the price per ton of reduced waste?'; while minimizing waste is costly, it is surely a lot cheaper than doing nothing.

MISSION OF THE SPV TEAM
The SPV team aims to establish commercially valuable applications and markets for all BPC by-product waste streams.

The objectives are three:
- avoid costs which are created due to the need for disposal, due to taxes levied on waste products, and due to other liabilities;
- create an income from selling these by-products in specific market segments and for specific applications;
- achieve environmental improvements for BPC through the reduction of waste and the avoidance of waste disposal operations.

The SPV team implements such activities in accordance with Responsible Care requirements and practices and aims to achieve these objectives by:
- recycling of products;
- internal re-use as raw materials or intermediate products;
- sales of products to third parties, with the aim of recycling or re-use as raw materials or as intermediate products.

SCOPE OF THE SPV ACTIVITIES

The main activities, which are related to SPV and which will be handled by the SPV team, consist of:
- plant surveys to identify SPV opportunities;
- marketing studies for recycling within specific markets;
- development of a data base for specific markets and products;
- development of contacts with selected companies for conditioning, upgrading and recycling of secondary products;
- developing or obtaining technology for the conditioning of secondary products prior to recycling;
- establish technigrams and Material Safety Data (MSD) sheets for the promotion and handling of secondary products. Obtaining toxicity and technical data and product approval;
- establish contracts for product conditioning by third parties;
- realizing sales of conditioned products in agreement with client's requirements;
- obtaining the necessary support from existing functional departments and groups, in order to realize the above in a cost effective way;
- maintaining awareness of environmental regulations and threats related to the product portfolio;

- developing a minimum of relevant procedures and systems to effectively support and control the activity.

Within this framework, the key activities are described here in more detail.

PLANT SURVEYS

The plant survey is a joint effort between selected plants and SPV team personnel to identify re-use, recovery and recycle opportunities. The survey may be conducted for a specific unit, product streams or for an entire site.

The plant survey encourages a company climate where each individual becomes alert to and seeks opportunities to convert disposal liabilities into saleable products.

PRODUCT PORTFOLIO

By building up a product portfolio of secondary products, the marketing approach becomes more effective in terms of:
- creating an attitute of product for a specific customer or application, rather than resigning to a 'waste product' situation;
- marketing by a dedicated team, rather than by the prime products marketing/sales team, which will be more inclined to treat secondary products as second priority;
- approaching market segments with a range of products, rather than with one single lower value product;
- easier networking on SPV opportunities and on potential customer contacts.

Once the business activity is established properly the SPV team will hand over to the relevant business area.

PRODUCT APPROVALS

Characterization of the secondary products and the preparatory work for product approvals and for product listing is one of the most critical and tedious parts of this activity. The success of it is entirely dependent on effective cooperation with R&D and Technical Service & Development (TS&D) departments, and with health, safety and environment specialists.

CRITICAL SUCCESS FACTORS

The successful operation of an SPV activity is highly dependent upon a number of critical success factors as shown in Table 10.1 overleaf.

TABLE 10.1
Secondary products valuization: critical success factors

Top management support

Learn from established programmes — for example, UCC

Find a customer need that matches with the need for a product outlet

Proven technology:
- for recovery of saleable products;
- for customer's process application.

Use simple and straightforward conditioning steps to adapt the by-product to customer requirements (see also Figure 10.1)

DEVELOPMENT OF THE ACTIVITY WITHIN A MULTINATIONAL ORGANIZATION

The SPV activity has started at one particular BPC site. The core SPV team is still located on that site and handles secondary products from all major BPC sites.

Key success factors for such organization are:

- work with an SPV 'champion' on each site;
- maintain good contacts with the relevant business divisions, both with technical and with commercial people;
- the company culture of network building and of sharing of best practices between sites and business divisions is a very powerful vehicle to carry this activity forward.

RESULTS AND BENEFITS

The success of SPV projects is measured as a sum of cost avoidance and variable contribution.

COST AVOIDANCE

The avoidance of real costs, related to the disposal of the by-product stream is reported for each SPV project. These costs include environmental taxes on waste products and handling and disposal costs.

VARIABLE CONTRIBUTION

The nett sales income for the secondary products is expressed as the sales income minus product conditioning and product handling costs.

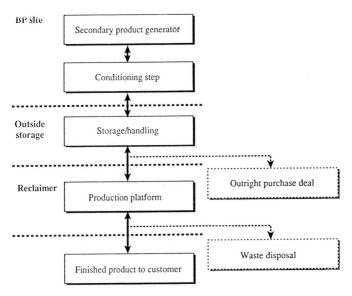

Figure 10.1 Recycling product flow.

1992 RESULTS

25 projects were started during 1992 and are summarized in Table 10.2.

TABLE 10.2
1992 Results

	Realized in 1992 ($)	Potential (*) ($)
Cost avoidance	0.18 m	> 1 m
Variable contribution	1.05 m	> 4 m
Total	1.23 m	> 5 m

(*) of the 1992 projects. In 1993 YTD 20 new projects have been started.

11. ALUMINIUM — A CASE FOR STRATEGIC ENVIRONMENTAL MANAGEMENT

Charles Duff

Total environmental impact studies can lead to new targets and priorities being set for the company, and this chapter describes how.

First, a few words about Norsk Hydro. We are Norway's largest company, with a turnover of £5.5 billion, which puts us at around number sixty in Europe. 40% of sales are derived from mineral fertilizers, of which we are the largest producer in the world. 30% of the business is in aluminium and magnesium, where we are respectively number one in Europe and number two in the world, and most of the remainder is in oil, gas and downstream products. We are large-scale operators in the North Sea, as well as in more exotic places; downstream activities include a chain of petrol stations in Scandinavia, but industrially more important is the manufacture of polyvinyl chloride (PVC) where, again, we are very substantial — European leaders in PVC compounds. Finally we are the largest producers of farmed Atlantic salmon in Europe.

All these activities we undertake in the UK in one form or another. We have some 2000 UK employees, operating on some 15 sites and generating £350 million sales, while we also import products worth another £400 million, notably natural gas which we sell direct to end users here.

I mentioned salmon farming, which of course is quite a small proportion of our total sales, to emphasize that all our activities have significant environmental impact. Fertilizers — are they associated with nitrate leaching into water courses? PVC — is it recyclable? Does it produce dioxins if it is incinerated? Aluminium — it's called the green metal because it is so easily recycled, but what about ore extraction and energy usage in its manufacture? And salmon farming — what about the visual impact of the farms in areas of outstanding natural beauty, and what about diseases and the release of aggressive antibiotics into the water?

Next, I should point out that most of our products have relatively low added value, so profitability depends on cost control, enhanced by economies of scale. We do have the great advantage of having hydroelectric power available to us — still virtually the only source of electricity in Norway, and we have our own stations — but even that is not free. Efficient and conservative use of energy and raw materials is critical to our success — more so than for operations leading to higher added value products.

As a final scene-setter, I should point out that in Norway — until very recently — we employed nearly half a percent of the total population. In proportion, that is the equivalent of British Telecom in the UK, and five times more than ICI! So we are highly visible, and deeply involved in the social, as well as the industrial and economic life of the country, This colours very strongly our attitude to our local responsibilities and our openness in communications.

So in summary we have a firm with a strong sense of social responsibility, with a product range which involves significant environmental impact, and with a cost structure which pushes us hard towards minimizing energy use and waste creation. This is what has led us to our current position in environmental management and reporting. I think the story of our activities in this field over the last few years is well known — an environmental report in Norway in response to adverse publicity, followed by a report in the UK, partly to raise awareness about Hydro and partly because we saw legislation moving in the direction of tighter standards and more reporting. Our 1990 UK report was outstandingly successful, largely because we included an external 'verification' (as the jargon now has it). We also set out to describe our products themselves — are they 'good' goods? — in response to some of the popularly held, though mistaken, views about them, such as those I noted earlier. These were early steps down the road towards total environmental impact, or life cycle analysis. It is work which we have taken much further since then, and it is what I would like to turn to now.

We are a vertically integrated producer of aluminium, going right back to ore extraction and forward to the manufacture of finished products. What I would like to show is that a careful review of the whole process from the point of view of total environmental impact may lead to some novel ideas about the disposition of resources, with the effect of both cutting the environmental impact and improving profitability.

Let's start with the production process. Bauxite is found these days mainly in Australia, Central America and Africa. 20% of it is located in areas of rain forest. Bauxite is aluminium oxide mixed with iron and other oxides so that it appears reddish-brown in colour. The bauxite is digested in caustic soda at elevated temperatures, and the aluminium oxide is precipitated out, leaving a sludge known as red mud. A tonne of this is produced for every tonne of alumina, the aluminium oxide. The alumina is calcined into the form of a white powder, which is then shipped to the aluminium smelter.

The smelters are located much closer to the eventual users, and notably in areas of low cost energy — in particular, those which have hydroelectric power. Nearly 60% of all smelting is done using hydropower, in places such as Canada and Switzerland as well as Norway. There is also a growing number of smelters in places such as the Middle East, where there are abundant supplies of natural gas at very low cost. Smelting is actually an electrolytic process; the molten metal collects in a bath-shaped cathode, and is then teemed off into a caster to form billets, for extrusion making, slabs for sheet and foil making, or ingots for castings and other products.

I would like to concentrate on the energy involved in these processes of primary metal production. The energy required for bauxite mining is relatively small: about 0.1 MWh per tonne of aluminium metal. Alumina extraction from the ore, however, has a substantial energy requirement — over 9 MWh per tonne of metal. In the areas where this process is carried out, the only practical form of energy is heavy fuel oil. Burning this to generate heat has the effect of producing 2.5 tonnes of carbon dioxide for every tonne of metal. Moving on to smelting (and ignoring the energy used in transporting the alumina to the smelter), the power used in smelting is around 20 MWh per tonne of metal. Now in Norway this energy is non-polluting hydropower. But the situation is that aluminium and oxygen are being separated, using electrodes which are made from carbon, at 950°C. Needless to say, the anode, at which the oxygen is released, burns away creating more carbon dioxide — another 2.0 tonnes per tonne of metal. So in total 4.5 tonnes of carbon dioxide are generated in producing a tonne of primary aluminium — the green metal!

It is clear that before aluminium gets out into the marketplace and has a chance to demonstrate its green credentials — low density, strength, resistance to corrosion, formability and so on — it does have a substantial, and adverse, environmental impact. As manufacturers, of course, we view at least some of these adverse effects as being a cost burden — if we could move to an inert material for the smelter anodes, for example, or improve energy utilization still further from the 30 MWh per tonne currently needed, then we would be saving money as well as reducing our environmental impact. These things, of course, we are doing. But there are other ways of looking at the problem.

Aluminium recycling is relatively straightforward. Because the material itself hardly degrades in use, aluminium products need only to be cleaned before they can be melted down and re-cast into the original primary form — billets, slabs, etc, ready for manufacture again, There is no quality deterioration and, most important of all, it consumes much less energy. Recycling uses only about 5% of the energy needed to produce the primary metal. We have a remelt unit ourselves in the UK — Hydro Aluminium Metals Ltd (HAML), in South Wales. It is an interesting case because its business is concentrated on the extrusion makers: HAML takes in crop ends, etc, from the extruders, melts them in a gas-fired furnace, and returns to them prime quality extrusion billet. The transaction is done on a tonne-for-tonne tolling basis, so there is no material element in the charge HAML makes for the service.

Other than labour, the company's only sizable costs are gas, and any new metal it has to purchase to replace material lost. We have written a case study about its development of environmental performance indicators as a management tool, and HAML is the only company in Europe to have been selected for the pilot studies of both BS 7750 and the European Community's eco-management and audit regulation.

So what about aluminium recycling? In the UK our consumption of aluminium is around 500,000 tonnes a year. Recycling of industrial products, such as car and building components, is quite high — 80% or more — and we, like other producers, are working with customers on 'design for de-manufacture' projects to simplify the task of recycling products made from our material still further. For consumer products, though — which really means drinks cans and foil — the recycling rate is only around 10%. (Can recycling is much higher in certain other countries, notably the USA and Sweden). And packaging accounts for around a quarter of all aluminium consumed in the UK.

We are starting to put two and two together about this. To go to the logical limit, would it be in our economic interests to put the marginal £1 million of expenditure not into further R&D on process improvement to save further decimal places of electricity, but into a marketing campaign to encourage more recycling among consumers? And when we consider that every ounce of alumina smelted in Western Europe is imported, and a burden on the balance of payments, is there not a case for seeking matching funding from the UK Government? Of course the situation is not quite so simple. There is a range of different alloys of aluminium and problems of segregation would at present provide some limitations, as would the cost of cleaning products from contaminants such as lacquer or grease — or congealed Coke. But the environmental benefits of cutting back on the extraction of non-renewable, finite supplies of bauxite, in environmentally sensitive parts of the world, and making great inroads into the energy consumed in primary manufacture, are attractive goals in themselves and could be very attractive financially as well. I do not have any breakthroughs to report to you yet, but we are looking at the situation very carefully.

This is what I mean by a strategic approach to environmental management. The tactics involve routine action at every stage of the production chain to cut down on energy and materials use and waste production, and as I have indicated there is clearly economic benefit in doing that. But the broader view offered by a total impact assessment is yielding interesting new lines of attack which could lead to significantly greater benefits. We are doing similar exercises on all our main products, and equally interesting results are being found.

12. ENVIRONMENTAL MANAGEMENT IN PACKAGING — RESPONDING TO THE DREAM
Lindsay Fortune

Packaging features on the list of concerns of most environmental groups. The concern may have little scientific basis but it has created a climate for legislative proposals which could have a major impact on the shape of the packaging industry and on the products available to users of packaging. While arguing the need for a sound scientific basis for legislation if real environmental benefit is to be achieved, this chapter reviews the current position and the implications of management of packaging production and use.

INTRODUCTION
Environmentalism has been said to be the major business issue for the 1990s. It is certainly significant for the packaging industry with implications for the products of the industry and the processes used to manufacture them.

The point can be made by considering just one current issue, the disposal of packaging after use.

Concern about the impact of human activity on the environment is high in most industrialized countries. As part of their agenda environmentalists have emphasized the finite nature of world resources. Any suggestion of waste triggers a negative response. When considering packaging, consumers judge resource use by the weight, or more commonly the volume, of material in their rubbish bin and so packaging has a negative image. Consumers have no guide to the hidden use of resources in the package — the energy to obtain the raw materials and to process them, the fuel to transport the packaging unfilled and filled — nor to the savings in product resources which have been achieved by the use of appropriate packaging.

But it is perception which stirs consumers and causes legislators to react. What follows considers public perceptions of packaging, some of the legislation being built on them, and some of the implications for industry if the legislation currently in place, or under discussion, becomes a permanent part of the body of law.

PERCEPTION

What do surveys tell us about the public perception of packaging?

- Used packaging is believed to be the major part of domestic waste.
- The use of packaging is sometimes considered 'unnecessary', frequently 'excessive'.
- The use of resources in packaging is thought to be growing.
- Packaging is said to be the cause of the litter problem.

There is general support for some action to reduce the use of packaging and particularly the amount of used packaging going to landfill. Environmental groups gain credibility by appearing to offer simple solutions, reducing science to a 'soundbite'.

'Packaging should be biodegradable'
'Packaging should be recyclable'
'Incineration is unsafe'
'Landfill is socially undesirable'

LEGISLATIVE ACTION

All around Europe, in North America and most other parts of the Western world, legislation is being proposed to deal with one or more of these perceived 'problems with packaging'. In Europe the focus is on promotion of recycling. The dream is that all packaging will be reused by refilling or by collection and recycling back to packaging.

The German Government has already introduced its Packaging Waste Ordinance which sets targets for the diversion of used packaging from landfill and sub-targets for the percentage of recovered packaging material which has to be recycled. By mid-1995, legislation requires 72% of all used glass, tinplate and aluminium and 64% of all other packaging to be delivered for recycling.

In the Netherlands, the packaging industry has signed a Covenant which targets 60% recycling of packaging materials by 2000.

In 1990 the UK Government also set an objective of recycling 50% of recyclable material from the domestic waste stream. This is effectively a 50% recycling target for packaging used for food and household goods.

The proposed EC Packaging Waste Directive is being debated currently. If it retains the key elements of the original proposal from the Commission, it will set a target for recovery of 90% of each material in the waste stream and the recycling of 60% of material recovered. The target would be expected to be approached progressively until approximately 2006. Other countries are proposing variants on these themes.

Although the legislation is proposed for environmental benefit, the objectives for the legislation are not always the same. In Germany the motivation is diversion of used packaging from landfill as not-in-my-backyard (NIMBY) pressures make it difficult for new landfill sites to be developed in that country. Motivation for the UK objective appears to be simply the belief in the virtue of recycling as such. The proposed EC Directive quotes several objectives:

- minimization of use of resources;
- diversion from landfill;
- encouragement of recycling;
- harmonization of measures throughout EC member states.

Legislation introduced to reduce the environmental impact of the use of packaging should be aimed at optimizing the use of resources in the broadest sense. Without such a specific objective, the reality is that much of the legislation being proposed around Europe in particular will not deliver nett environmental benefit but will result in an increased use of resources.

While not all proposed national legislation will generate environmental benefit, some will distort trade patterns, affect the shape of the packaging industry and create advantages for domestic producers in the countries in which the legislation is introduced.

There is no doubt that an EC Directive on this issue is needed to ensure consistency in approach through Europe. Targets need to be challenging and environmentally sensible. Those in the proposed Directive are not environmentally sound.

IMPLICATIONS

What then are the implications if European governments press on with their present line of legislation on the management of used packaging?

All industry will need to review its packaging systems
- There will be growth in the use of standardized, reusable packaging, particularly for bulk transport.
- There will be reduction in the use of secondary packaging — for example, a carton to pack another package such as a toothpaste tube.
- There will be growth in the use of refillable containers. The detergent industry

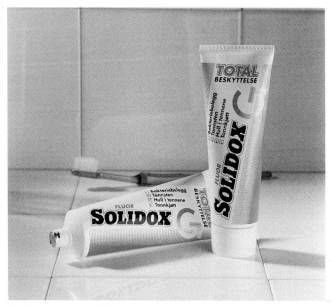

Toothpaste tubes which stand on the display shelf and thus eliminate need for the secondary packaging of an outer carton has been one response to environmental pressures. (Courtesy of Courtaulds plc.)

is already marketing concentrated product in lightweight packs for dilution by the consumer and storage in a larger container retained in the home.

If recycling targets by material are introduced throughout Europe, there will be a decline in the use of laminated structures and plastic and paper packaging in general, except for paper board and corrugated containers.

- Development of a successful recycling system requires three elements; an appropriate collection and sorting infrastructure, facilities to clean and reprocess recovered materials and outlets for recycled materials. For plastics and paper and particularly for laminates, there are major limitations on the reuse of recycled material unless the specification is known. Once wastes are mixed, only limited separation of plastic and paper to specification is practicable. Furthermore, waste may be contaminated by inks, coatings, pigments or product remains. The recycling process causes some deterioration to material properties. The end result is that insufficient outlets are available now for all materials which

could be collected. The future size of the plastics and paper packaging industries would be determined by development of outlets for recovered plastics in non-packaging applications.
- From the domestic waste stream, only recovery of the bulkier rigid plastic containers such as bottles appears to be environmentally justifiable.
- A study by a working party from the flexible packaging industry[1] demonstrated that, for the UK, recycling plastic film from industrial use is environmentally sound and is expected to satisfy all available end use markets.
- Plastic waste collection in Germany in 1993 is expected to reach a level of 200,000 tonnes per annum against a current processing and reuse capacity of 60,000 tonnes. At present Germany is exporting surpluses and undermining the recycling industries in other countries. The progressive application of the German regulations will increase collection to three and a half times the current rate by the end of 1995. As no major technological breakthrough can be envisaged which can be in place by then, either the plastics industry must reduce production, space be found for a plastic waste mountain or the law be modified. Industry will have to reconsider selection of packaging materials.
- Reduction in availability of plastic and paper packaging materials will mean a reversion to the packaging range of 20–30 years ago.
- As plastic packaging currently packages about 50% of all food products, the reduced availability of plastics will reduce the variety of product offer.

There will be an increase in the overall weight of packaging and in the amount of total resources used in packaging.
- An example from Germany (see Table 12.1 overleaf) demonstrates that switching from lightweight laminated materials to recyclable materials which are recovered and recycled in accordance with the German regulations will result in a much greater weight of material going to landfill than the total weight of the laminated material.
- The increased weight of packaging requires lorries to transport the packaged product with associated congestion and pollution.

REALITY VS PERCEPTION

AMOUNT OF PACKAGING WASTE
Packaging waste is frequently cited as a major component of waste streams. In

TABLE 12.1
Coffee packaging in Germany

Pack type	Flexible laminate (tonnes)	Tinplate container (tonnes)	Glass jar (tonnes)
Packaging required	11,280 tonnes	122,200 tonnes	473,572 tonnes
Treatment of used packaging as per ordinance:			
recovery (*0%)	9024 tonnes	97,760 tonnes	378,858 tonnes
recycle (90% glass and tinplate)	7219 tonnes	87,984 tonnes	340,972 tonnes
Balance for disposal	4061 tonnes	34,216 tonnes	132,600 tonnes

Source: German Coffee Association[2]

truth it is a minor component. In the UK each resident generates close to 9 tonnes of waste per annum. The bulk of this (78%) arises from the activities of the agricultural, mining, quarrying and construction industries which provide the essential raw materials. A further 18% is produced on our behalf by industry and commerce. Households generate directly about 4% of waste through the dustbin.

Regular analyses of the UK dustbin by Warren Spring Laboratory has shown that packaging comprises around 25–30% by weight. Thus in the UK packaging represents about 1% of the solid waste requiring disposal. Analysis of the content of the dustbin is given in Table 12.2.

ROLE OF PACKAGING
In developed countries, the use of packaging is so well established that its vital role in containing, identifying and protecting products and in informing consumers tends to be taken for granted. The benefits of packaging are best understood by those who do not have a modern packaging and distribution infrastructure. In Europe it is estimated that only 2–3% of food goes to waste before it reaches the consumer compared with 30–50% in countries which do not have Europe's distribution system.

TRENDS IN THE USE OF PACKAGING
Accurate statistics on the actual weight of packaging used in any country are

very difficult to find but it is certain that in developed countries the use of packaging is growing less than growth in GDP and in some countries declining in absolute terms. Studies by Dr Harvey Alter in the USA showed that population grew by 18% between 1970 and 1986 and municipal solid waste by 25%. Within the municipal waste stream, packaging grew by 9% over the period compared with 40% growth in reading materials and 69% growth in other paper waste.

Similar trends exist in the UK also where the weight of paper in waste is not less than the weight of packaging materials and is growing more rapidly.

LITTER

Packaging is frequently associated in the public mind with the problem of litter. However, packaging itself does not cause litter. Littering is a social problem and needs to be dealt with as such by the development of appropriate social responsibility.

TABLE 12.2
Composition of UK domestic (dustbin) waste by weight

	Average analysis (%)	Approximate packaging content (%)
Combustible materials		
Paper/board	33.7	5.2
Rigid plastic	5.9	3.4
Flexible plastics	5.7	5.5
Textiles	2.2	—
Other	8.5	—
Non-combustible materials		
Glass	9.2	7.8
Metal	7.2	6.1
Other	1.5	—
Compostable materials		
Food/garden waste	20.1	—
Dust	6.0	—
	100	28.0

Source: Warren Spring Laboratory, composite of 60 surveys

BIODEGRADABILITY
During the 1980s biodegradability was promoted by some environmentalists as a desirable contribution to reducing the environmental impact of packaging. On closer analysis most have backed away from this position recognizing that biodegradability is an unpredictable mechanism with initiation difficult to control. The introduction of features to trigger biodegradability artificially add to the expense of materials and limit access to other disposal options such as recycling. Furthermore changing materials which would otherwise be stable in landfill into degradable materials which could generate landfill gas could well be environmentally most undesirable.

RESPONDING TO THE DREAM
The environmentally sustainable approach of the packaging industry must be to optimize package design to prevent loss of the product being packaged while minimizing the use of resource in the composition and production of packaging materials.

The evolving science of life cycle analysis will play an increasingly significant role in developing environmentally-optimized package design.

Appropriate disposal of the waste must form part of that total programme of optimization. Each packaging material has its own specific contribution to make to reducing the impact of packaging waste on the environment. Some materials are best suited to recycling, others contribute to an overall reduction in the use of resources in packaging, still others are ideal for incineration with energy recovery.

The EC strategy for waste management defined three broad guidelines for dealing with wastes in order to utilize resources better and reduce volumes sent to landfill:

- prevention (source minimization);
- reuse or recycling;
- reduction of volume or harmfulness.

The UK packaging industry already has a demonstrable record of lightweighting to minimize the use of resources. This point is demonstrated in Table 12.3.

To minimize the environmental impact of packaging and optimize the processing of packaging waste, a comprehensive waste management system

TABLE 12.3
Weight of typical packages used in UK

Approximate date	Milk bottle	Yoghurt pot	Food can	Drinks can	Garbage bag
1960	397 g	12 g	69 g	91 g	—
1970	340 g	9 g	69 g	91 g	50 g
1980	245 g	7 g	58 g	20 g	41 g
1990	245 g	5 g	57 g	17 g	24 g

Source: INCPEN

suitable for processing all packaging materials needs to be developed. There are five options to be considered as means of recovering value from packaging waste:

- reuse;
- material recycling;
- feedstock recycling;
- energy recovery;
- composting.

Table 12.4 (see overleaf) shows, however, there is no disposal option which is universally suitable for all packaging materials. A recent life cycle study by the Packaging and Industrial Films Association (PIFA)/Flexible Packaging Association (FPA)/Oriented Polypropylene Manufacturers Association (OPMA) working group[3] demonstrated through a life cycle analysis that recycling was a valid option for processing unprinted polyethylene film from industrial use (pallet wrapping) back to carrier bag film, but not for processing thin printed film recovered from domestic waste when additional cleaning was required plus extra film thickness to compensate for contamination. The energy comparison is shown in Table 12.5 overleaf. Washing and recycling contaminated film consumes as much energy as production of virgin granule.

The group considered that recycling of plastic film waste in the UK should focus on industrial waste while the best route for disposal of plastics in domestic waste was incineration to recover the energy content of the materials and convert it to electricity or local heating. The group noted the UK lagged well

TABLE 12.4
Options for processing used packaging

	Reuse	Material recycle	Feedstock recycling	Energy recovery	Compost
Glass	**	**			
Steel	*	**			
Aluminium		**		*	
Rigid plastics	*	*	*	**	
Flexible plastics	*	*	*	**	
Cellulose/ cellulose acetate				**	**
Wood	**	*		**	
Composites		*	*	**	
Paper		*		**	*
Board	*	**		**	*

Key: * Limited scope   ** Universally applicable

TABLE 12.5
Energy comparison in recovery of polyethylene film

	Virgin granule GJ/tonne	Unprinted industrial waste GJ/tonne	Contaminated domestic waste GL/tonne
Collection	—	0.3	1–2
Sorting	—	—	1
Wash/recycle	—	23–30	60–90
De-ink	—	—	25
Total energy use	92	23–30	87–117
Production efficiency loss	0	4–5	22–30
Total energy in feedstock	92	27–35	109–147

TABLE 12.6
Incineration of waste in Europe

Country	% incineration capacity with energy recovery	Total disposal by incineration
Luxembourg	100	75
Switzerland	90	59
Belgium	62	54
Denmark	100	48
Sweden	100	47
France	75	42
Germany (FRG)	100	36
Netherlands	97	35
Norway	100	22
Italy	72	16
United Kingdom	29	8
Spain	79	6

Source: TNO report for PWMI, 1992

behind the rest of Europe in the use of incineration in waste disposal and in the number of incinerators equipped to utilize the energy released (Table 12.6).

Support for the case for increased investment in incineration in UK has recently come from the Royal Commission on Environmental Pollution[4].

CONCLUSION

The packaging industry faces uncertain times. In the name of the environment, legislation is threatened which would reshape substantially the industry, making forward planning difficult. The problem is compounded by different action being taken in different countries, and even regions of countries, in response to lobbying on local priorities from local sectoral groups.

In Europe, Germany is implementing legislation for the management of used packaging waste which in effect results in German consumers subsidising the raw materials for certain sectors of the German packaging industry and threatens recycling industries in other parts of Europe while achieving no obvious environmental benefit.

What is needed now is European legislation which encourages the establishment of a waste management infrastructure to enable the optimization of disposal of all waste. Any targets set for used packaging management should be environmentally sound and not based on an assumption that recycling should be the preferred disposal option for all materials. The proposed EC Packaging Waste Directive would introduce conflict between the EC strategic objectives of minimization and recycling of packaging materials.

One problem is the absence of agreed criteria for measuring environmental impact. The science of life cycle analysis offers prospect of a tool which will enable various packaging options to be evaluated, once the methodology is developed and widely adopted by industry.

The packaging industry needs to be more active in encouraging the adoption of resource-efficient packaging and to continue to support development of programmes to reprocess used packaging. It needs to be more effective in publicizing progress, both historic and future, in lightweighting and in assisting development of waste management options.

Users of packaging must expect changes and should be reconsidering all existing packaging and introducing changes which make ecological sense, like:

- returnable packaging;
- refillable packaging;
- elimination of secondary packaging.

In the UK there is need for installation of modern waste incineration capacity with energy recovery in order to optimize waste disposal.

From the UK and elsewhere in Europe there is need for some action to persuade Germany to deal with the packaging waste which they collect and not export it to other countries to the detriment of those countries' recycling activities.

REFERENCES IN CHAPTER 12

1. *Management of waste plastic packaging films*, June 1991, PIFA/FPA/OPMA.
2. Private communication, Deutscher Kaffee-Verband to Dr K. Topfer, 3 April 1992.
3. *Valorization of used flexible packaging*, January 1993, PIFA/FPA/OPMA.
4. *RCEP 17th Report, Incineration of waste*, 1003.